STUDIES IN MODERN HISTORY

General Editor: **J. C. D. Clark**, *Fellow of All Souls College, Oxford*

Editorial Board

T. H. Breen, *William Smith Mason Professor of History, Northwestern University*

François Furet, *Professor of History, Ecole des Hautes Etudes en Sciences Sociales, Paris*

Peter Laslett, *Fellow of Trinity College, Cambridge*

Geoffrey Parker, *Professor of History, Yale University*

J. G. A. Peacock, *Professor of History, The Johns Hopkins University*

Hagen Schulze, *Professor of History, Universität der Bundeswehr, München*

Norman Stone, *Professor of Modern History, University of Oxford*

Gordon Wood, *Professor of History, Brown University*

The recent proliferation of controversy in many areas of modern history has had common causes. The revision of assumptions and orthodoxies, always professed as the role of scholarship in each generation but seldom really attempted, has increasingly become a reality. Historians previously unused to debating their major premises have been confronted by fundamental challenges to their subjects – the reconceptualisation of familiar issues and the revision of accepted chronological, geographical and cultural frameworks have characterised much of the best recent research. Increasingly, too, areas of scholarship have passed through this phase of conflict and recasting, and works of synthesis are now emerging in idioms which incorporate new perspectives on old areas of study. This series is designed to accommodate, encourage and promote books which embody the latest thinking in this idiom. The series aims to publish bold, innovative statements in British, European and American history since the Reformation and it will pay particular attention to the writings and insights of younger scholars on both sides of the Atlantic.

Giambattista Vico

Imagination and Historical Knowledge

Cecilia Miller
Assistant Professor of European Intellectual History
Wesleyan University, Connecticut

palgrave

Published by
PALGRAVE
Houndmills, Basingstoke, Hampshire RG21 6XS and
175 Fifth Avenue, New York, N.Y. 10010
Companies and representatives throughout the world

PALGRAVE is the new global academic imprint of
St. Martin's Press LLC Scholarly and Reference Division and
Palgrave Publishers Ltd (formerly Macmillan Press Ltd).

ISBN 0–333–55153–2

This book is printed on paper suitable for recycling and
made from fully managed and sustained forest sources.

A catalogue record for this book is available
from the British Library.

Transferred to digital printing 2001

For Gordon DesBrisay

Contents

Acknowledgements

I begin by thanking Isaiah Berlin, who supervised my Oxford D.Phil. thesis on Vico, for his enthusiasm and support throughout my research. Conversations with Donald Verene, Leon Pompa, Laurence Brockliss, John Robertson and Joseph Mali spurred me on in my studies. I am indebted to Giorgio Tagliacozzo for putting me in touch with many Vico scholars in Italy and Britain.

Quentin Skinner and Patrick Gardiner examined my thesis and I thank them for their comments. Donald Kelley, Perez Zagorin, Dan Terkla and Giuseppe Mazzotta gave me detailed criticism of the manuscript.

I am grateful for the funds I received during my time in Oxford from the Overseas Research Scheme, the Oxford University Scholarship and Balliol College.

The Centro di Studi Vichiani made me very welcome in Naples and provided me with much needed materials. Lessico Intellettuale Europeo in Rome furnished me with draft versions of its forthcoming concordances to several of Vico's writings.

Librarians at the Taylorian Library and the Bodleian Library (Oxford), the British Library, the London Library, the Biblioteca Nazionale «Vittorio Emanuele III» di Napoli, the Butler Library (Columbia), the Beinecke Library (Yale), the Widener Library, the Houghton Library and the Law Library (Harvard) and the Library of Congress assisted me.

Eleanor Wood, Jerri-Lyn Scofield and Julia Perkins gave me excellent help with the manuscript. James O'Hara and Christopher Parslow corrected my Latin cheerfully. Neil Parekh helped with the index.

The Society of Fellows at Columbia University provided me with a very congenial venue for work on this project as a Mellon Postdoctoral Research Fellow. Thank you to my colleagues at Wesleyan University for their comments on my book.

I appreciate the help I received from my editors, Giovanna Davitti and Anthony Grahame, at Macmillan.

My parents, J. Melvin and Wanda (Howard), and my siblings, Kyle and Celesta, read the manuscript. I thank them for their unfailing support throughout my research and for giving me a love of books and of debating ideas.

Thank you to William Fike.

Note on the Texts

Unless otherwise indicated the Latin and Italian references are to the Nicolini edition of Vico's writings. There are three exceptions. A new edition of the orations has been produced by the Centro di Studi Vichiani, *Institutiones oratoriae* has been translated into Italian by Giuliano Crifó. The 1730 edition of *La scienza nuova* was not reprinted in its entirety by Nicolini; in this case, references are made either to the extracts by Nicolini or to the page numbers in the 1730 edition itself. References to *La scienza nuova* (1725, 1730 and 1744) are made either by Nicolini's paragraph divisions or by Vico's own divisions. The autobiography is referred to by page number in the Nicolini edition. All other divisions are Vico's own.

There are several excellent English translations of Vico – by Bergin and Fisch (the 1744 edition and the autobiography), Pompa (extracts from the 1725 edition and *De antiquissima italorum sapientia* as well as others), and Gianturco (*De nostri temporis studiorum ratione*). The translations of the Orations by Giorgio A. Pinton will be published shortly by Cornell University Press. Translations quoted from these works are by the scholars listed above. All other translations are my own. Mainly due to Vico's idiosyncratic spelling, the original spelling and punctuation have been retained in quotations.

Select List of Vico's Writings

Affetti di un disperato	1693
Inaugural Orations	1699, 1700, 1701, 1704, 1705, 1707
De nostri temporis studiorum ratione	1709
De antiquissima italorum sapientia	1710
De rebus gestis Antonj Caraphaei	1716
Il diritto universale	1720–22
La scienza nuova prima	1725
Vita di Giambattista Vico scritta da sé medesimo	1725, 1728
La scienza nuova	1730
'La pratica'	1731
De mente heroica	1732
Institutiones oratoriae	1741
La scienza nuova	1744

List of Abbreviations

Inaugural Orations (1699–1707)	Orations
Vita di Giambattista Vico scritta da sé medesimo (1725, 1728)	Autobiography
La scienza nuova (1725)	*1725*
La scienza nuova (1730)	*1730*
La scienza nuova (1744)	*1744*
Section of *La scienza nuova*, unpublished in Vico's lifetime, entitled *'La pratica della scienza nuova'*	*'La pratica'*

Introduction

Vico's fundamental importance in the history of European ideas lies in his strong anti-Cartesian, anti-French and anti-Enlightenment views. In an age in which intellectuals congratulated themselves on their modern (which was to say, rational) approach to life, Giambattista Vico (1668–1744) stressed the nonrational elements in man – in particular, imagination – as well as social and civic relationships, none of them easily reducible to the scientific theories so popular in his time. It is well known that the chief reason Vico gave for his anti-Cartesian stance was that man could not fully know the natural world (science) as it was made by God, but that human history was largely, if not entirely, comprehensible precisely because it was man-made. Yet it is important to note that although Vico turned his back on Cartesian rationalism, he nevertheless applied rational methods to the subject René Descartes (1596–1650) despised – history.

Vico's main effort was an attempt to discern, if not a pattern, at least some otherwise unobtainable glimpses into the past. His most important means were language, mythology and rites of religion, all of which were to be used as tools in this exploration into the unrecorded periods of human history. As alternatives to written records, they gave access to the past civilisations that left no historical documents. Vico's unique contribution was to view myths not as false statements about reality or fanciful versions of past events, but rather as embodiments of early outlooks and beliefs; similarly, he was not as concerned with cycles in history, *per se*, but in their use as an instrument to investigate the development of cultures.

Vico's notion of the development of the human mind and the corresponding development of social institutions has not been analysed sufficiently. Vico saw the anomalies and idiosyncrasies of mankind as natural parts of the creative spirit, which could not be forced into any kind of methodological straitjacket. It was Vico's belief in the nonrational aspect of human nature which separated his work from that of his contemporaries.

Vico's ideas were indeed out of step with the intellectual climate, not only in late seventeenth- and early eighteenth-century Naples, but in Europe as a whole; certainly, they did not fit neatly into the Enlightenment. Yet it was precisely these ideas which constituted his unique contribution to the history of ideas and which laid the groundwork for subsequent inquiries into the philosophy of history. Vico's concern not only with patterns of history, but, more importantly, with the issue of changes in human nature and in society

1

is, if anything, of more pertinent interest today than at the time they were written. This book attempts to provide a reinterpretation of several of the theoretical foundations of Vico studies. It endeavours to counter the belief in Vico's epistemological break (1710), at which point he supposedly became suddenly and violently anti-Cartesian. The anti-Cartesian tone of his later works is not in doubt, but there are (in this writer's view) elements of his most original ideas concerning imagination and historical knowledge in his pre-1710 writings.

One reason the accepted view has remained unchallenged for so long is that many previous studies have restricted themselves to an analysis of the third and final version of Vico's best-known work, *La scienza nuova (The New Science*, 1744). Little effort has been made to trace the development of his ideas, not only in the critical first (1725) and almost unknown second (1730) editions of *La scienza nuova*, but also throughout his autobiography and earlier theoretical writings. The first six orations and the 1725 and 1730 editions of *La scienza nuova* deserve particular attention here, since these works remain relatively untouched by Vico scholars, who have generally ignored the strong statements that these works contain concerning imagination in regard to social and historical development. At the heart of this research, then, is the attempt not only to analyse Vico's theories of history as reflected not only by his well-known cyclical view of history, but also to give a systematic analysis, based on all his theoretical works, which it is hoped will establish the crucial position of his profound insights regarding imagination and human creativity in relation to historical knowledge.

*   *   *

In order to comprehend Vico's profound statements regarding language, imagination and historical knowledge, it is crucial to have an understanding of his own particular, rather idiosyncratic, vocabulary. This requires an analysis not only of all three versions of *La scienza nuova* (1725, 1730, 1744) and his autobiography (1725, 1728), but also of *Il diritto universale (Universal Law*, 1720–22) and all of his earlier theoretical writings (1699–1710, 1719). There are several themes to which he returned time after time in his writings: uniformity of ideas, discussed both in terms of human nature and common sense (*sensus communis* [Latin] and *senso comune* [Italian]), primitive wisdom (*sapienza volgare)*, the idea of society, social structures and his new critical art. These concepts commanded Vico's interest precisely because he was persuaded that they could inform him concerning past societies. The history which Vico sought to explore had little to do with chronologies of rulers or particular events, except as they could be used as evidence of an individual society's philosophy of life.

Vico's work is indeed distinguished from that of most of his contemporaries by his strong, anti-Cartesian stance. It has been previously accepted that in 1710 Vico experienced a dramatic epistemological break in this regard. That Vico's writings were profoundly anti-Cartesian after this date is not at all in doubt. However even in his earliest writings, the 1699–1707 orations, elements of his most inventive notions of imagination (*phantasia* [Latin] and *fantasia* [Italian]) and historical knowledge are to be found, although they admittedly appear side-by-side with more traditional Cartesian notions, particularly those regarding the importance of mathematics and physics. Nevertheless an intellectual history of Vico's thought cannot ignore these early attempts to grasp the importance of these two essential concepts simply because they are not yet presented in a purified form.

According to Vico there were seven main aspects to his work. It was the elucidation of these principles, shared by all societies, which he believed to be his particular task. The first was 'a rational civil theology of divine providence' ('*una teologia civile ragionata della provvedenza*'). Arguments still continue regarding the exact relationship of Vico's 'divine providence' to Adam Smith's (1723–90) 'invisible hand' and Georg Wilhelm Friedrich Hegel's (1770–1831) 'cunning of reason'. But it can be stated with some assurance that Vico's divine providence had little to do with any rational pattern or the increase or development of reason. Divine providence was for Vico the providential path which shapes the unforeseen consequences of life for the benefit of those involved.

Secondly, Vico wrote of a 'philosophy of authority' ('*una filosofia dell'autoritá*'), which was always discussed in relation to the Law of the Twelve Tables. This insight is one of Vico's most profound. He argued that the Law of the Twelve Tables – he used the same type of argument regarding the authorship of the *Iliad* and the *Odyssey* – was not gleaned from Greek law but was the result of the collective wisdom of the ancient Italian peoples. Here was an early case of exposure of an obvious anachronism, a fanciful juxtaposition of elements in culture which belong to quite different stages of development – the ancient Italians not only did not derive their law code from Solon (638?–?559 B.C.); they could not have. The implications of this statement went far beyond the identification of the origins of even such an important law code (or of epic poetry). Vico was convinced that the 'order of ideas must follow the order of institutions' ('*L'ordine dell'idee dee procedere secondo l'ordine delle cose.*'). Thus law and literature were prime examples of a particular social group's creativity, which was not dependent on timeless, Platonic patterns.

A 'history of human ideas' (*'una storia d'umane idee'*) was the third main idea on Vico's list. He considered primitive poetry, and equally primitive religion and law, to be created by the same faculty and that all expressed the way of thinking and feeling of a specific society. These ideas determined the character of a society, and were the object of Vico's method of historical reconstruction.

Vico's fourth point was 'a philosophical criticism' (*'una critica filosofica'*) which developed from the history of ideas. He maintained it was possible to enter by means of the imagination into ancient, even pre-literate societies, by means of a critical examination of social usage, thereby reconstructing certain forms of social behaviour. His historical method was not all-forgiving; rather it offered a means to examine critically cultures quite diverse from one's own. Vico did not moralise; instead his was an attempt to gain isolated glimpses of these past societies. His warnings regarding anachronisms were the basis on which such an investigation could be carried out.

Next was Vico's concept of an 'ideal, eternal history traversed in time by the histories of all nations' (*'una storia ideal eterna sopra la quale corrano in tempo le storie di tutte le nazioni'*), which he declared to be an eternal truth. Here we must note Vico's tripartite historical cycles which have traditionally received more attention than any other aspect of his work. The cycles were not an unimportant part of his thought, but they certainly were not the most original.

Sixth, according to Vico, was natural law (*'il diritto natural delle genti'*), which was composed of all the concrete, as opposed to transcendental, aspects of human society such as the origins of religions, languages, customs, governments and other social creations. For Vico the study of these institutions was the correct means to gain insight concerning a past society. Vico regarded natural law, divine providence and his 'ideal, eternal history' as the principles upon which human history was worked out. Yet his main focus was not on these abstract concepts as much as on social institutions, the study of which he assumed was the only sure means to apprehend human history.

Vico's crucial, final point had to do with the principles of universal history (*'i principi della storia universale'*). Since later men were unable to enter into the imaginations of the first men, Vico offered *fantasia* as the method, the means, to overcome this barrier and to reconstruct the thinking of these early peoples. The term *fantasia* was also used by Vico to describe the creative efforts of these same peoples (in the form of poetry, religion, law, or any other social institution) and at the same time the resultant culture of that particular social group. Vico's emphasis was always on the historical and social dimensions. *Fantasia* was the means, the *scienza nuova*, which allowed historical reconstruction and thus provided the historical knowledge which Vico sought.

Vico's primary concern was with the ways of thinking and feeling, the mentalities of distant, often all but forgotten societies. Most of his discussions focused on the first stages of a primitive social group. Myths and early poetry were seen by Vico as virtually identical for the purposes of later historical analysis with the outlook, the *Weltanschauung*, of these particular social groups. It was *fantasia* which Vico identified as the peculiar, involuntary force which created early poetry. And it was *fantasia* which comprised the culture, the particular contributions of any given society. It was also *fantasia* which one must make use of in order to enter into the world view of these peoples so far removed from oneself, both chronologically and culturally.

Nevertheless it must be clearly stated that Vico showed little interest in using his historical method, *fantasia*, himself. Most of his references were to ancient Rome, an exceptionally well-documented society. He made passing references to North American Indians and the ancient Chinese, for example, but his only sustained discussion of an early society was of the Greeks. This observation will hardly come as a surprise to the reader of Vico, nor does it downgrade the importance of *fantasia* and historical knowledge. Vico was always much more interested in the theoretical underpinnings of any concept than its practical application.

*   *   *

Vico studies have gone through several definite phases. Vico was generally ignored outside of Naples in his lifetime. There were only a few exceptions. *De antiquissima italorum sapientia (On the Wisdom of the Ancient Italians)* was reviewed unfavourably in the *Giornale de' letterati d'Italia (Journal of the Scholars of Italy*, 1711). Jean Le Clerc (1657–1736) wrote a review which was more of a summary, although without an understanding of the quite radical implications of *Il diritto universale* (1720–22) in the 1722 volume of the last edition of his *Bibliothèque ancienne et moderne (Ancient and Modern Library*, 1714–22). Vico wrote his autobiography for publication, at the request of a Venetian nobleman, to serve as a model for future intellectual autobiographies. And *La scienza nuova prima (The First New Science*, 1725) was reviewed in Leipzig in 1729. But these isolated contacts brought Vico no lasting connection with the wider intellectual world. He was further restricted by his inability to read any modern European language except Italian, which was unusual for Neapolitan academics of his generation. His only other language was Latin.

In the generations following Vico, Antonio Genovesi (1713–69), Gaetano Filangieri (1752–88) and other economic thinkers of the Kingdom of Naples often referred to him as the source of their inspiration. The actual connection between Vico and these later economic theorists was quite slight, but their desire to identify themselves with him was in itself significant. Vincenzo Cuoco (1770–1823) was the best known of those who had to flee Naples at the time of the Neapolitan Revolution of 1799 and who subsequently introduced Vico's works to France and northern Italy. Over the past two-and-a-half centuries Catholics, Romantics, Italian nationalists, German historicists, communists and fascists – to name just a few groups – have all claimed Vico as their spiritual father or son. Isolated references to Vico, by Karl Marx (1818–1883), for example, or the integration of Vico into later fictional works, with James Joyce (1882–1941) as the prime example, have coloured our perception of his work much more than would have been possible with a better-known thinker. But even though these groups and thinkers were generally not addressing the Vico to be read in the texts, their fascination with Vichian topics is itself compelling. An intellectual history of the tradition of Vico studies and the diverse interpretations and misinterpretations of his thought remains to be written.

Jules Michelet (1798–1874) was the first of Vico's great interpreters. His Romantic interpretation of Vico idealised men in the state of nature, and stressed the notion of humanity struggling to raise itself above the pressure of external forces. Although interesting on its own terms, Michelet's interpretation of Vico no longer adds much that is new to an understanding of Vico. Vico was analysed by Idealist thinkers Giovanni Gentile (1875–1944) and Benedetto Croce (1866–1952). Croce's powerful and extremely persuasive pronouncements regarding Vico, which referred constantly to the transcendental aspects of history, are still a force to be reckoned with in Vico studies, both in the Italian-speaking world and beyond. R. G. Collingwood's (1889–1943) translations of Croce made these views accessible to the English-speaking world as early as 1913. For this reason, and because of the major impact of Collingwood's own brief writings on Vico, it is necessary for the modern scholar to take great care that assumptions regarding Vico can be traced back to Vico himself and did not originate with Croce or Collingwood. At present it is Isaiah Berlin's interpretation, in which he deals with Vico's stress on the non-rational elements in man and Vico's original methods of examining history, which dominates the field both inside and outside Italy. Michelet, Croce and Collingwood, and Berlin represent three extremes in Vico studies – the Romantic, the Idealist and the Liberal.

Much important work has been done relatively recently in Vico studies by Adams, Badaloni, Battistini, Donzelli, Fassò, Fisch, Fubini, Giarrizzo, Gianturco, Haddock, Momigliano, Mooney, Piovani, Pompa, Rossi, Said, Verene and Zagorin to name just a few. The tricentenary of Vico's birth in 1968 gave rise to a profusion of conferences, special collections and volumes dedicated to the Neapolitan thinker. Two institutes have been founded – the Institute for Vico Studies in New York and Atlanta and the Centro di Studi Vichiani in Naples. Such an upsurge of interest in Vico is not to be disparaged, for it must be assumed that it represents a very real curiosity regarding Vico's own concepts. Nevertheless much work on Vico remains to be done. The lack of an English translation of his complete works is perhaps the most serious issue. In addition, there is a need for a collection of essential articles on Vico, which remain scattered in often obscure journals. There is also a need to view Vico's thought in an interdisciplinary manner. His eighteenth-century views do not fit neatly into the modern understanding of history, philosophy, politics, language and literature.

*   *   *

This book, then, is an effort to explicate and analyse what I believe to be the two central themes of Vico's thought – imagination and historical knowledge. For the most part, this study has been based on a textual approach to Vico's theoretical works. His work sustains detailed scrutiny very well. For this reason it is maintained that one need not necessarily study his writings in conjunction with those of a better known thinker or intellectual movement, although such studies, for example those by Badaloni, most certainly have their own advantages. Thus little attempt has been made here to discuss at length either Vico's sources or his extremely disparate followers.

Perhaps surprisingly, there follows no lengthy discussion of Croce's interpretation of Vico at any length. Croce's views are so pervasive that an analysis of *Vico senza Croce* (Vico without Croce) seemed to be in order. However far we have moved from Croce's Idealism, his discussion of Vico and mythology still maintains its relevance today. The works of Fausto Nicolini (1879–1965), the great editor of Vico, form the backbone of any Vichian study. As this book follows most closely in the tradition of Collingwood and Berlin, its points of departure from their interpretations are spelled out explicitly throughout.

This work is distinguished from that of Leon Pompa by its stress on imagination, which, it will be argued, is by no means a marginal aspect of Vico's thought. I argue that Vico cannot be placed neatly into the tradition of Western philosophy. To do so creates a danger that Vico might appear to be a second-rate philosopher, and the diversity and richness of his thought and the originality of both his subjects and his views might be ignored. Bruce Haddock and Giovanni Giarrizzo maintain, in their own ways, that Vico was a political thinker. This book will argue that Vico was uninterested in political theory and political structures; that which one might wish to regard as a philosophy of politics was for Vico an investigation into the social relations of particular cultures. This work also takes a different route from that of Donald Verene on knowledge (although there is no disagreement on the primacy of imagination) as this study analyses the relationship of *fantasia*, in the various senses it was used by Vico – to culture, society, language and history.

It is the purpose here to establish two main points. The first is to show that an analysis of all the versions of *La scienza nuova*, as well as Vico's earlier theoretical texts, can provide a means to gain a more fully-rounded appreciation of his thought. Special emphasis has been placed on the second version of *La scienza nuova* (1730) and the first six Inaugural Orations (1699–1707), precisely because of the information they contain concerning imagination. This textual study has led to the contention that Vico's epistemological break was neither so sudden nor unaccountable as has been previously assumed. Secondly and most significantly is the argument that the concepts of *fantasia* and historical knowledge were the central themes in Vico's thought, and that they require an appraisal not only of early language but of the development of the very *idea* of society. The aim is to demonstrate that it was these notions of imagination, language and historical consciousness which constituted Vico's unique contribution to the history of ideas and which laid the groundwork for subsequent inquiries into the philosophy of history.

1 Vico's Intellectual Development

1. Vico's Orations (1699, 1700, 1701, 1704, 1705, 1707) and His Supposed Epistemological Break (1710)

Virtually all recent work on Vico is based on the premise that there was an epistemological break in his thought in 1710, at which point he became suddenly and dramatically anti-Cartesian.[1] One reason for this is that most previous studies have restricted themselves to an analysis of the third and final version of Vico's best known work, *La scienza nuova* (1744). Little effort has been made to trace the development of his ideas not only in the important first (1725) and almost unknown second (1730) editions of *La scienza nuova*, but also throughout his autobiographical and earlier theoretical writings. The exception to this practice is the attention given to *De antiquissima italorum sapientia*, the publication of which in 1710 is held to mark his supposed intellectual conversion. Quite to the contrary, however, it can be argued that no such major shift in his thought occurred at any point. Yet this is not to deny Vico's anti-Cartesian stance. The year 1710 was not a dramatic turning point in Vico's intellectual development; rather, it was the year that he wrote the first of his works which was to receive significant attention.

Without a doubt Vico was violently anti-Cartesian by the time he published *De antiquissima italorum sapientia* (1710) and certainly his autobiography (1725, 1728).[2] But elements of his most original views, those regarding the importance of imagination and historical knowledge, are to be found in the first six orations as well, admittedly side-by-side with praise for Descartes. The ambivalence of Vico's views in these orations should be viewed in its historical and intellectual contexts: praise of mathematics and physics would not have been surprising to Vico's readers. The point at which Vico dropped his Cartesian references altogether and began to actively criticise the French philosopher in his later works is well known.

However we should not ignore his first six orations simply because in them he praised Descartes and adopted the mathematical method. It must be remembered that Vico praised the mathematical method in 1720–22. As late as 1744 he again examined the merits and demerits of various scientific approaches, long after his alleged epistemological break. Vico became anti-Cartesian, but never anti-scientific, even though his own interests veered towards the human and social sciences. More intriguing than his comments on science or Descartes are his first discussions of imagination and historical knowledge. This argument is not only about texts, but about ideas, ideas which should not be dismissed because they are mixed with unoriginal material. Indeed all of Vico's works contain very well-known issues, most particularly the 1744 edition of *La scienza nuova*, to which most Vico scholars confine themselves.

Vico's opposition to Descartes had at least as much to do with the scientific approach, which developed for the most part after the French philosopher's death and which bore his name, as with Descartes himself. This distinction is critical since it helps somewhat to explain the ambivalence in Vico's writings from 1699–1710, in which he both praised Descartes and stressed the importance of imagination and history. Vico's incorporation of Cartesian themes into his earliest theoretical works was not unusual for the time, and thus it is no excuse for ignoring the peculiar ideas of his own which were present there as well. It can be argued further that Descartes deserves to rank, even if in opposition, along with Vico's four acknowledged *autori* – Plato (427?–347 B.C.), Tacitus (55?–after 117 A.D.), Francis Bacon (1561–1626) and Hugo Grotius (1583–1645).[3]

There is no denying that the epistemological break retains its fascination on many levels. It makes Vico an honorary Modern by detaching him from his own cultural and intellectual background. It gives Vico an experience that ranks with Descartes's famous night in the winter of 1619 in a 'stove-heated' room in Bavaria.[4] It seemingly explains Vico's switch from a discussion of the nature of man in the first three orations (I–III) to that of the development of civil societies in the next three (IV–VI), although even this shift occurred as early as 1704.[5] Further, the scientific references returned in full force in his later works; thus there was not the complete turn from scientific matters that one would expect from such an about-face.

Hence there was no radical shift in 1710, since Vico's discussions in the first years of the eighteenth century on human nature and society demonstrate a continuum of the very themes which he was to make so famous. If there was a break, it was with the discussion of his method, his new science, his critical art, in *Il diritto universale* written in the years 1720–22; but even this approach was firmly embedded in the early works.[6] The radical break in Vico's thought that many commentators have for so long accepted simply is not to be found in his writings. The reality which commanded Vico's interest throughout all of his writings was social relations. Cartesian thought was shed in a natural and gradual process, as he turned his attention to language and societies. In particular, scientific Cartesian thought was dropped as

Vico turned his attention more and more towards the relationships among social groups. There is a parallel here with Vico's own discussion of the development of societies, in which transitions from one stage to the next were natural and gradual, not dramatic and abrupt.

As Vico's anti-Cartesian views are not in doubt, when reading Descartes one is struck more forcefully by the parallels with Vico than with the contrasts. Repeatedly the same issues were addressed (imagination, memory, will, good or common sense) even though the conclusions are contradictory.[7] Even less well known are the topics upon which Vico and Descartes agreed. Each felt that it was necessary to begin intellectual endeavour afresh. Both believed they had found a new method which would explain and unify all subjects, and they shared (along with Bacon) an unmitigated contempt for Scholasticism.[8] Vico was not, however, at all grateful to Descartes for helping to loosen the grip of Scholasticism on academic life.

Both Vico and Descartes prized common sense, and a childlike awareness of the world over the scientific (or any other) theories taught from books; even from their diametrically opposed standpoints on the relative importance of the arts and sciences, they each designed new systems, new approaches to human knowledge.[9] Descartes wrote that it was more effective to re-do the whole scheme of human knowledge, rather than to revise isolated aspects, giving the example of town planning (that it is easier to plan a new city than to renovate an existing one), but without much discussion of the inherent problems of implementing such a scheme.[10] Descartes's desire was to reform human learning by showing that all disciplines were parts of a single science.[11] Both Vico and Descartes wanted to unify and to provide a method for the explanation and study of human knowledge.

Nevertheless it is not difficult to find Descartes's famous denunciation of the arts – Part I, paragraphs 8 and 9 of the *Discours de la méthode* – and the accessibility of this passage is no doubt one reason that the surprising parallels between the two thinkers have been almost entirely ignored.[12] It is not, however, the intention here simply to catalogue little known points in common between the two thinkers in order to claim a direct link between them. Rather it is instructive to realize that the goals and outlook of the two were quite similar. This view is most important in terms of Vico's supposed epistemological break. Although Descartes and Vico developed conflicting systems, they started with many of the same goals and dealt with similar issues. This helps to explain Vico's early admiration for Descartes and how he was able to mix Cartesian concepts with his own for so many years before he dropped the Cartesian aspects altogether.

Vico's views were in many cases the mirror image of those of Descartes – the analogy could be made of a child who rebels against his parents by adopting diametrically opposed political and social views. This idea goes a long way towards explaining Vico's rather rabid anti-Cartesian statements in the Autobiography and his eagerness to disassociate even his youthful self from the Cartesian spirit so prevalent in eighteenth-century Naples, where in the academies – if not at the university – it was considered a mark of distinction to say that one understood Descartes.[13]

There is no doubt that at the time Vico wrote his autobiography he had become violently anti-Cartesian.[14] As Yvon Belavel's summary of Vico's views indicates:

[Descartes] . . . has slipped some fictions into his *Discours* . . . ; he does not acknowledge his readings . . . ; in his "greed for glory" he has planned, in opposition to the *Metaphysics* of Aristotle, to establish "his empire in the cathedral schools" . . . ; . . . he has become the leader of a sect . . . ; his method is sterile; some of his followers are quacks . . . ; the final result of his philosophy is fatalism, and so on.[15]

Vico not only made personal attacks on Descartes's character, calling him overly ambitious for glory, for example, but more importantly he could not forgive Descartes for the scorn the French philosopher had poured on the study of language, orators, historians and poets – the very elements which together comprised the essence of Vico's approach.[16] In any case Vico's anti-Cartesian stance was a means by which to demonstrate his break with his contemporaries and identify himself with the Ancients. Hence his much-vaunted epistemological break of 1710 was a break not only from science (or even from law) to history, but was also a split from the current scientific trends in Naples and Europe. One might argue that Vico's major attack was as much on scientific Cartesianism as on Descartes himself, which would have pleased Descartes, for he pleaded that his readers believe only him, via the texts, not what was said about him.[17]

According to the French philosopher, important subjects must be analysed in an unbroken manner, beginning with the simplest and most evident truths; in this manner Descartes dealt with transitions in nature and in history.[18] He maintained that the whole of human knowledge consists of a distinct perception of the way in which these simple ideas combine to build up other objects. Thus he, like Vico, dealt with transitions in history as a gradual process.[19] And it was Descartes, not Vico, who wrote that one must not go beyond what one understands intuitively.[20] Belavel wrote:

No longer is imagination – for Vico any more than Descartes – a simple auxiliary of the intellect, which the union of soul and body renders useful.

We should, instead, envisage it as the primal, positive power of seizing analogies and similarities; without that power, chance would never result in creation . . . [21]

This is not to say that Vico and Descartes were closely aligned even on non-scientific imagination. According to Descartes intuition was 'an unclouded conception of an attentive mind and springs from the light of *reason* alone' ('*mentis purae & attentae non dubiam conceptum, qui à fola rationis luce nafcitur*').[22] Descartes contended that imagination is most intense when the brain is disturbed, when the true is linked with the real and the false with the fantastic.[23] In addition he contended that ideas did not come via the senses.[24] Although Vico and Descartes both desired to reorder the division and examination of intellectual endeavours and addressed many of the same subjects – imagination, memory, will and common sense – their approaches were sharply divergent.

For his part Vico was not at all concerned with imagination or ideas in the Cartesian sense.[25] There is no discussion in Vico of when ideas may be present, the difference between perception and ideas, or even a clear differentiation between memory and imagination, much less any concern regarding the distinction between the mind and the body. For Vico imagination was not irrational but nonrational.

The utility of doubt so important to Descartes was completely missing in Vico. Descartes desired to find just one thing that was certain, indubitable, whereas Vico was completely unconcerned with this issue. For Vico clear and distinct ideas were only abstractions of the human mind. He asserted:

. . . ac proinde nostra clara ac distincta mentis idea, nedum ceterum verorum, sed mentis ipsius criterium esse non possit: quia, dum se mens cognoscit, no facit, et quia non facit, nescit genus seu modum, quo se cognoscit.	Accordingly, our clear and distinct idea of the mind cannot be a criterion of the mind itself, still less of other truths. For while the mind perceives itself it does not make itself, and because it does not know the genus or mode by which it perceives itself.[26]

In the same work, *De antiquissima italorum sapientia*, Vico made his famous declaration that only by making something can we hope to understand it. In the same manner we can only hope to understand events in the past, according to Vico, if we re-make them by means of our own imagination.[27]

Descartes was troubled by the paradox that we cannot doubt our existence without existing while we doubt. Vico called this 'Descartes's deceitful demon' ('*genio fallaci Carthesii*'); for Vico the famous Cartesian tag would be an affirmation: I think, therefore *we*, as a society, mankind, are.[28] According to Vico his *scienza nuova* was possible if we could recover the principles from the modifications of *our same* human mind.[29]

The concept of God was used by Descartes as a means to discuss limitless will and intelligence. Descartes appealed to God's veracity to bridge the epistemological gap between belief and certain knowledge.[30] Yet one of his more famous critics, Antoine Arnauld (1612–1694), was one of the first of many to attack the circularity of this argument. According to Arnauld, Descartes used clear and distinct ideas to prove the existence of God, while at the same time appealing to the veracity of God to guarantee clear and distinct ideas – this is not dissimilar to Vico's treatment of natural law and sacred religion.[31] For both Descartes and Vico the truths they sought to identify and order were distinct from God, and theology only entered into their writings as it furnished a basis of certainty. Descartes believed one could only understand the world through reason, whereas Vico considered reason to offer only a partial solution, and Vico argued that imagination was a much more profound method.

For Vico the only creations worthy of sustained consideration were social institutions; he was not at all concerned (following Oration VI) with the physical world or even the composition of human nature in the manner of Plato, Thomas Aquinas (1225?–1274), Bacon or Descartes, although Vico continued his discussion of human nature in his later works in a very different form as *senso comune*.[32] His earlier works (the Orations, *De nostri temporis studiorum ratione*, *De antiquissima italorum sapientia* and the autobiography), listed ways to feed the imagination: start children in school late, where they should read both epic poetry and national histories and study the quantitative sciences, in particular, geometry, since he considered geometry to be a tool in the formation of an inventive mind.[33] But if *De nostri temporis studiorum ratione* with its attack on the Modern, French approach was a leap ahead, then *De antiquissima italorum sapientia* (with the exception of the first chapter of the first book on *verum* and *factum*) was, in some respects, a step backwards, for in it Vico discussed at length mathematics, physics, motion and extension. Thus in 1709 in *De nostri temporis studiorum ratione* (and as late as 1744) he espoused the virtues of the sciences.[34] But this was possible, according to Vico, because one could recognise the usefulness of mathematics and physics without being Cartesian. Hence Vico professed a limited approval of the geometrical (and hence Cartesian) method because it developed a taste for order and, most importantly, because it developed the imagination. Vico sought to counter what he viewed as the extremes of Cartesianism by re-establishing the dignity of the science of man.[35]

Indeed Vico was very much out of his depth as a reviewer of Cartesianism. Even as a student he never demonstrated any aptitude for the sciences; his background was in the classics and law. There is even some doubt as to to Vico's familiarity with Descartes, especially with the *Discours de la méthode*, since it was written not in Latin but in French, which Vico could not read. Nevertheless Vichian criticism of Descartes, if not of the highest standard, is of interest, if only because it indicates some of Vico's weak points as well as his own intellectual priorities.[36]

The importance to Vico of the belief that he developed in an intellectual vacuum, and of his related feelings concerning Descartes, Isaac Newton (1642–1727) and John Locke (1632–1704), no doubt had much more to do with his feelings of exclusion (both real and imagined) from the intellectual, academic and social hierarchies of Naples, let alone Europe. Vico was not so much in rebellion against his European intellectual heritage as he was in reaction to the university in Naples and the decline of its fortunes, his own lowly position there, and what he perceived to be the generally poor intellectual standard at the beginning of the eighteenth century.[37] It is very probable that if Vico had been more successful professionally, he would have felt that the historical methods available were all that were necessary, and would never have developed his own views on imagination and historical knowledge.

If Vico had accepted that he would not be famous in his own time then this is another reason that he would have written on generalised topics which he would have hoped to be of interest to future readers.[38] Whereas Vico turned to the Ancients for inspiration, Descartes looked to the past masters to see what was left for investigation.[39] Vico lashed out against the Moderns and their scientific approach because he felt that they would not allow a study of cultures and other topics (now termed social sciences) which cannot be discussed in scientific terms. For all these reasons Vico considered himself an Ancient, yet his views are often of more interest to us than to his Modern (or even Ancient) contemporaries.

Thus Vico had always been anti-Cartesian in outlook, even if he himself did not fully recognize it until later in his life. A recognition that there was no dramatic epistemological break in Vico's thought is important not only in terms of the surprising unity of Vico's work, it also establishes the critical position of his early works. A short survey of Vico's first six orations clearly demonstrates not only the presence of what we now consider to be Vichian themes, but also lengthy discussions of the same. It is accepted that in Vico's first three orations (1699, 1700, and 1701), he dealt with human nature and the development of the mind, and in the next three (1704, 1705, and 1707), he discussed the usefulness and, indeed, the necessity for society to foster both a love of literature and a well-rounded educational system, embracing both the arts and the sciences.[40] But beyond these common topics of the time, the orations were the first expression of Vico's own ideas regarding society and history.

As early as the first oration, *Suam ipsius cognitionem ad omnem doctrinarum orbem brevi absolvendum maximo cuique esse incitamento* (*Knowledge of oneself is for everyone the greatest incentive to acquire the universe of learning in the shortest possible time*) delivered on 18 October 1699, Vico outlined many of the most important themes which would play major roles in his work over the next four decades: *phantasia* (Latin for imagination), the faculty which conceived images; Greek and Roman gods as early expressions of *phantasia*; the uselessness of music or the performing and creative arts in developing the mind; the central role of memory; philology as an historical method (his concern with the history of words – philology and etymology – was due to his obsession with the history of culture); *ingenium* (promptness, acuity, dedication, capacity, inventiveness and immediacy) as the guide in all creative enterprises; and knowledge, which he considered to be closely related to the will. It was here that he first made the critical point that it was by means of *phantasia* that one could reclaim both the great and sublime aspects of the past.[41] As early as Oration I (1699) we find his initial statement concerning imagination, *phantasia*, as the creative faculty.

Vis vero illa rerum imagines conformandi, quae dicitur «*phantasia*», dum novas formas gignit et procreat, divinitatem profecto originis asserit et confirmat. Haec finxit maiorum minorumque gentium deos, haec finxit heroas, haec rerum formas modo vertit, modo componit, modo secernit; haec res maxime remotissimas ob oculos ponit, dissitas complectitur, inaccessas superat, abstrusas aperit, per invias viam munit. At quanta et quam incredibili velocitate!	Truly, the power that fashions the images of things, which is called phantasy, at the same time that it originates and produces new forms, reveals and confirms its own divine origin. It was this that imagined the gods of all the major and minor nations; it was this that imagined the heroes; it is this that now differentiates the the forms of things, sometimes separating them, at other times mixing them together. It is phantasy that makes present to our eyes lands that are very far away, that unites those things that are

separated, that overcomes the
inaccessible, that discloses what
is hidden and builds a road
through trackless places. And it
does all this with unbelievable
swiftness![42]

In the first oration (after some very slightly veiled criticism of the Rector of the University of Naples for asking him so late to give this speech) Vico asserted that all men have the desire for self-knowledge, but only educated people have the ability and opportunity to recognise and then act on this compulsion. For this reason he considered students to be naturally attracted to learning. Professors, he argued, have an obligation to teach without bias, in order to satisfy their students's needs. (He always wrote as if it were possible to teach or write without any bias, neglecting to notice that his own views formed a particular outlook, or prejudice, of their own.) This notion of an innate desire for knowledge was an early parallel of Vico's own tenet that societies naturally preserve records of their past and seek ways to decode such artefacts. Vico's use of self-knowledge (*'suam ipsius cognitionem'*)[43] had no modern psychoanalytical overtones; nor did it mean an acceptance of one's own mental and physical limitations. Rather it was recognition of an inherent desire to learn. For Vico the possibility of becoming wise depended essentially on our will, our determination. It was cultural knowledge, knowledge of the past of one's own society and of the the natural world, which he sought.

The desire to comprehend the past was for Vico a basic human need. Here as elsewhere in his work, the discussion of the development of laws and human institutions was curiously amoral. Vico did not (as John Stuart Mill, 1806–1873, was to argue in the following century) regard laws as the product of intellect and virtue, nor of modern corruption grafted upon barbarism.[44] Instead Vico viewed the development of these specific human institutions and the subsequent artefacts – laws in codified form – as a natural, and always gradual, process. Laws, according to Vico, always reflected the spirit of the entire society concerned, not just its elite.

The second oration was entitled *Hostem hosti infensiorem infestioremque quam stultum sibi esse neminem* (*There is no enemy more dangerous and treacherous to its adversary than the fool against himself*, 18 October 1700). His discussion of eternal models and eternal order in this work referred to his faith in the Christian religion, not an acceptance of Platonic universals. He stated that ferocity, bestiality and war should be avoided or discouraged, and in their place he offered the students the satisfaction which comes from intense study and the freedom generated by wisdom. In general this oration is less original and offered his listeners such platitudes as the need to reproduce in oneself the order seen in nature.[45]

However the tension between reason and passion within man, a critical issue for students, according to Vico in this oration, is of more than passing interest. For Vico unrestrained passions were the weapons of fools (*stultorum arma sunt effraenes animi affectus*)[46] and thus reason should be supreme over the passions. Yet he did not state that it should have the dominant role over *phantasia*. For *phantasia* was not identical with the passions for Vico. It could be instinctive and it certainly was non-rational, but *phantasia* was assigned to the category of the spirit (not the emotions), which included also *sensus communis*.

In the third oration, *A litteraria societate omnem malam fraudem abesse oportere, si nos vera non simulata, solida non vana eruditione ornatos esse studeamus* (*If we would study to manifest true, not feigned, and solid, not empty, erudition, the republic of letters must be rid of every deceit,* 18 October 1701).[47] He argued that free will, a 'magnificent gift from God', was also responsible for much violence. Free will is abused when actions are taken which are not good for society as a whole or for the environment, nature. For Vico the possibility of society was built on reciprocal trust (for, he mentioned, even criminals obey the laws of their nefarious organizations) and the proper use of human reason. The distinction he later drew between the will and reason is in this work rather muddled. One explanation might be that in early societies he would not have considered the difference to be very great between them. In this work he also praised Aristotle for his philosophy of customs; this was exactly what Vico himself was to create in later years.

This oration is essential as it stresses the bonds which tie men together in society. He spoke of an innate desire in man to associate with others – demonstrating his confidence in the natural sociability of man. In an age dominated by the Hobbesian view of warring primitives in the state of nature, the Lockean notion of early men being unable to use their natural reason in the earliest stages of society, and later Jean-Jacques Rousseau's (1712–78) noble savages forced into society by families and farming (albeit, Rousseau's work was yet to be written), Vico insisted not on man's natural good nature but on his innate desire to associate with others. Vico obviously followed Aristotle (348–322 B.C.) on this point, but he differed from Aristotle in viewing this inclination not as a desire for political control and economic security but as an inner drive without which (and thus without language, law and all social creations) man could not be fully human.

In the fourth oration, *Si quis ex literarum studiis maximas utilitates easque semper cum honestate coniunctas percipere velit, is rei publicae seu communi civium bono erudiatur (If one wishes to gain the greatest benefit from the study of the liberal arts, and these always conjoined with honor, let him be educated for the good of the republic which is the common good of the citizenry,* 18 October 1704), he reiterated the definition of *phantasia* as the faculty of forming images of things and he again mentioned *ingenium*, which he described as exuberant.[48] In this he stated for the first time his oft repeated belief that imagination was strongest in the young, be they individuals, communities or whole cultures.[49] He praised students, who, after long nights of study, would come through pouring rain to attend the lectures at the university. He desired to encourage this drive and curiosity in the young. Yet a warning note was sounded regarding *phantasia*, which he equated with youthful enthusiasm. While it should be nurtured in the young, Vico declared, it should not dominate in intellectual endeavours. Indeed at this stage not only *phantasia* but the senses and even reason are all dismissed as insufficient for the young ever fully to grasp the arts and sciences. Vico's emphasis on history is more comprehensible in this respect, for he felt it was the one subject of which we could gain the most complete knowledge.

Proper conduct in both education and society was of particular interest to him in this oration. In terms of early society he argued that political positions were created because of the need to help the community. At every stage of a civilization, he maintained citizenship to be useful, because it breeds feelings of piety and respect, presumably in general, not just for the country involved. For Vico a university liberal arts education was necessary not only for the individual, but also for the state, since graduates could then be employed to work for the government. This was an old Neapolitan tradition – the university in Naples was founded in 1224 by the Hohenstaufen Frederick II for the explicit purpose of training civil servants. For this reason there was a general sentiment in Naples that education paid for by the state was worthwhile. The purpose of this speech was certainly to encourage the students, but his reasons were not entirely altruistic, because, as Vico himself stated, society cares for the young so that later they will care for society.

The fifth oration, *Res publicas tum maxime belli gloria inclytas et rerum imperio potentas, quum maxime literis floruerunt (Nations have been most celebrated in glory for battles and have obtained the greatest political power when they excelled in letters,* 18 October 1705) is of the least lasting value. It discusses war and honour, with specific examples given of each.[50] Vico's answer to the debate regarding the relative merits of military and literary glory, was that they complement each other. Vico taught that literary glory would follow military glory and that the same people would not do both. He believed scholars to be as productive as soldiers in their work for the state. Leadership qualities are stressed in this oration as well as in the previous one. His preference for a strong state which would foster study in the humanities is absolutely clear here. Although of far less merit than the other orations (as indicated by his classical references, a sure sign that he is not dealing with a topic in an original manner) it does indicate his interest in the military, in the concept of heroism, and in the symbolism with which it is invested.

The final oration, *Corruptae hominum naturae cognito ad universum ingenuarum artium scientiarumque orbem absolvendum invitat, ac rectum, facilem ac perpetuum in iis addiscendis ordinem exponit* (*The knowledge of the corrupt nature of man invites us to study the entire universe of liberal arts and sciences and sets for the correct method by which to learn them*, 18 October 1707), is the most exciting of the orations. He began it by discussing wisdom and the will.[51] The role of wisdom he saw as threefold:

Tria ipsissima sapientiae officia: eloquentia stultorum ferociam cicurare, prudentia eos ab errore deducere, virtute de iis bene mereri, atque eo pacto pro se quemque sedulo humanam adiuvare societatem.	Three are the very duties of wisdom – with eloquence to tame the impetuousness of the fools, with prudence to lead them out of error, with virtue toward them to earn their good will, and in these ways, each according to his ability, to foster with zeal the society of man.[52]

Vico declared that the benefits of learning always accrued to society. This theme was to be taken up enthusiastically in his later works, when he discussed the formation of society. In this work he stated that languages were the most powerful means for setting up human societies; for language would have been essential to form the type of associations he discussed in Oration III. But the motivating force was a self-centered love, a desire for advancement, usually associated with the notions of Bernard Mandeville (1670?–1733) and Smith.

Also in this oration, far from deprecating mathematics, Vico cited it as a means to develop the imagination, and he observed that this sort of study should be required for the young, when their imaginations were strongest. He went so far as to state that imagination should be sculpted into young minds at a very tender age so that it could not then be erased.[53] He recognised the immense power of *phantasia* and sought to harness it, so that later in life it would overwhelm reason. This potential conflict between reason and *phantasia* should not be viewed as a denouncement of *phantasia* by Vico, as much as an indication of the importance he placed on a proper balance between the two.

One last theme in this work must be mentioned: his belief that the foolish do not have the ability to distinguish the truth – a view that was to receive pride of place in *La scienza nuova*.[54] This theme had earlier been discussed by Descartes:

En quoy il n'eft pas vrayfemblable que tous fe trompent; mais plutoft cela tefmoigne que la puiffance se bien iuger, & distinguer le vray d'avec le faux, qui eft proprement ce qu'on nomme le bons fens ou la raison, eft naturellement efgale en tous les hommes; . . .

It is unlikely that this is an error on their part; it seems rather to be evidence in support of the view that the power of forming a good judgement and of distinguishing the true from the false, which is properly speaking what is called Good Sense or Reason, is by nature equal in all men.[55]

Les plus grandes ames font capables des plus grans vices, auffy bien que des plus grandes vertus; . . .

The greatest minds are capable of the greatest vices as well as of the greatest virtues.[56]

This final point, unfortunately, was never addressed by Vico, nor is there any discussion of restraints on the ruler in Vico.

One of the reasons that these early works have been ignored is due to the claim that there was no method in them. Issues of civil society and cultural development were discussed, but it has generally been assumed that Vico gave no hints as to how to approach them. The reason may be because *phantasia* has not been previously recognised as his means of obtaining historical knowledge, and thus it has been assumed that it was only when Vico used the terms *scienza nuova* and *arte critica* that he had a system in mind.

It is indeed amazing that parallels could exist between Vico and Descartes, the latter believing that the 'gracefulness of fables make one imagine many events as possible which in reality are not so' ('*Outre que les font imaginer plufieurs eunemens comme poffibles qui ne le font point*').[57] Descartes tried to purge himself of all beliefs without a rational basis, writing that 'we must be particularly careful not to let our reason go on holiday while we are examining the truth of any matter' ('*vt illis confifa ratio, etiamfi quodammodo ferietur ab ipfius illationis evidenti & attenda confideratione*').[58] Vico was more interested in man's emotions than his rational development, for the limitations on the subjects which could be pursued by rational means were all too clear to Vico. Although there is in his writings a combination of scientific with mythical, poetic truth, particularly in *De nostri temporis studiorum ratione*, it is not a sophisticated admixture.[59] In Vico the poetic passages soar, while the scientific ones seem leaden. Vico's brilliance was in recognising that poetry could be used in a sophisticated way in order to find out about the past, by means of the historical imagination. Hence these *universali fantastici* (imaginative universals) embedded in *sapienza volgare* (common wisdom, in the form of the myths) are revealed by means of conscious reflection upon them.[60] Myths and fables were used by Vico as the basis for re-constructing a world view of primitive peoples. According to Vico analyses of these early myths and fables must be done as much as possible from the vantage points of these same peoples. Vico warned against the dangers of an overly sophisticated or nationalistic approach, but once he had issued these warnings he was quite willing to use any technique or approach to revive these lost mentalities.

Although Vico listed Plato as the first of his *quattro autori*, it has for some time been recognised that the most profound influence came in the form of neo-Platonism, not from reading Plato. Vico's introduction to the concept of imagination as well as law in the neo-Platonic writers he read at Vatolla has not been previously recognised.[61] Gianfrancesco Pico della Mirandola's (1463–94) treatise *De imaginatione* (published posthumously in 1501) is for the most part a condemnation of the evils of imagination.[62] Yet it now seems clear that it was from Pico that Vico gained his incurable fascination with the topic of imagination. No doubt Vico's grounding in neo-Platonic thought helped shield him from the pursuit of scientific, Cartesian ideas. This was due to the contradictory implication of neo-Platonism which placed rational knowledge infinitely above the reach of the human mind.[63] Vico's adoption of Platonism rather than Cartesianism (for, as has been established, imagination was an essential faculty to Descartes, but for him it never threatened the supremacy of reason) in turn overcame the neo-Platonic problem that human knowledge appears as a feeble vestige of divine knowledge, by means of Vico's elimination of any attempt to deal with nature or the sciences.[64] Nevertheless man's inability ever to recover the totality of human history would be an issue in this regard – human knowledge of history would be but a feeble vestige of divine knowledge. Still Vico considered his critical art, his new science, enabled historical reconstruction to be a fruitful (if not altogether complete) and necessary endeavour.

No ideal society was ever presented in Vico's writings as either a model or a goal because he was not concerned with social change for the future but with an analysis of the taming of primitive man.[65] Both Descartes and Vico recognized that the people were the source of power for a state and a society but neither was particularly interested in methods of governing the people. Vico offered no new ideas regarding the civic virtues that children should absorb, but a modern variation of the republican virtues, the philosophy of man expounded by Petrarch (1304–74), Lorenzo Valla (c. 1400–57), Marsilio Ficino (1433–99), Pietro Pomponazzi (1462–1525) and especially Pico can be assumed to have been Vico's model.[66] Yet these civic virtues receive only one mention in Vico, whereas the benefits of education are almost omnipresent in his work, especially in the orations and his autobiography.[67] In this way Vico's *scienza nuova* may be viewed as an educational programme. His critical art had to be mastered and then applied, in order not only to understand the past properly but also to maintain the present order.

Vico was scathing about cultures which did not maintain their dominance and were overtaken by other societies on the rise. He never explained just why growth was natural, but not decline. For a thinker who so often discussed world history and the great civilisations of the past, it seems rather inconsistent that he attached a moral value and adopted a judgemental stance concerning societies which had lost their vigour and were either gently or rapidly declining. The explanation for this anomaly seems at first to be straightforward – that Vico began to take a personal interest in the maintenance of advanced societies, as this study had a personal relevance for him and his own time. However it seems more plausible that the reason that Vico genuinely feared the decline of society had nothing to do with his own situation, or with issues of right or wrong. For a thinker so occupied with the concepts of imagination and human creativity (both necessary for the maintenance of society), the death of a fully functioning civilisation meant the end of invention, and for Vico, without man's social creation, there was nothing left worthy of investigation.

Much later thinkers were concerned with a time when, again according to Mill, 'the law came to be like the costume of a full-grown man who had never put off the clothes made for him when he first went to school'.[68] For Vico, this sort of society would truly be finished. Law was all-important to him in its role as an indicator of a society's structural growth. But Vico never placed law above history; he contended that law could only be understood in relation to a specific historical context. At one point Vico maintained that all universal science was summed up in the legal sciences. In *Il diritto universale*, which more than any other of his works dealt with specific groups and events, Vico's philosophy of law was already based on the science of human cultures, which could be examined by means of *De principiis humanitatis* (*The Principles of Humanity*; an early, working title for *La scienza nuova*).[69] Vico did not view law as radically different from literature or written histories. Previously he would have noted this approach in some of the neo-Platonic writers he read at Vatolla who discussed imagination, literature and the law in the same sections of a single treatise.

At the heart of Vico's message was his belief that history itself was a human artefact and thus comprehensible by means of a careful examination of human institutions such as laws, religion and customs. Unlike most thinkers of his time, Vico saw the comprehension of history as an end in itself, not, in the words of Laurence Brockliss, as 'a means to criticise existing legal systems (which had departed from the unwritten law common to all men) or as a way of justifying *raison d'état* (in which present necessity meant that written law was different from the universal law of man)'.[70] Historical knowledge had precedence over all other types of knowledge for Vico, especially over scientific thought – theological knowledge was always exempted from his discussions. It was nature that we cannot know, for Vico, science and mathematics are what we do know. According to Vico historical knowledge was sadly neglected for the simple reason that there was not a widespread belief in its usefulness. Vico's aim was to provide a method which would enable one to re-make these cultures of antiquity, in spite of the intervening years. Even though Vico devised his own chronologies, his major effort was to break down the chronologies of ancient narratives and to discover and order their common themes, in order to understand the universality of myths, thus providing a basis for the study of individual cultures.

There is no evidence that Vico was personally concerned about reconstructing past societies; rather it was the theoretical underpinnings of such an attempt which entranced him. This was not necessarily the case because he considered the abstract aspects of a subject (for example, preliterate societies) to be inherently more interesting than a study of well-known ancient texts, but because these past cultures were completely unsolved puzzles, and thus, among other reasons, much more of a challenge to him. Once revealed they could illuminate the texts which he so revered.

Vico did not construct a 'tree of knowledge' – along the lines of Bacon or Denis Diderot (1713–84) and Jean d'Alembert (1717–83) – because his aim was not to divide but to unify what he considered to be the legitimate fields of human knowledge.[71] Vico paid virtually no attention to Bacon's last category of reason, which included the faculties of the mind and government as well as the sciences and medicine, although for Bacon and certainly for

Descartes reason was always the superior faculty to imagination. Ultimately Vico left both his favourite *quattro autori* – Plato, Tacitus, Bacon and Hugo Grotius – as well as his critique of Descartes, instead stressing the categories of collective memory and imagination, and he limited his definition and analysis of knowledge to what was humanly knowable.

Far from a dramatic epistemological break there was a slow shift in Vico's thought, and, arguably, an ambivalence throughout the whole of his writings regarding the sciences and the study of man. For Vico gave qualified approval to some Cartesian theories, while at the same time advocating a balanced approach in the university curriculum which would include the sciences as well as the arts. Despite Vico's later anti-Cartesian statements, and examples of these attitudes can be found in his work as early as 1699 and as late as 1744 – the belief in Vico's sudden epistemological break in 1709–10 is too extreme.

2. *Verum ipsum factum*

The above discussion does not detract from the importance of *verum ipsum factum* (that the true is what is made) (*De antiquissima italorum sapientia*, I,1). Far from a dead-end approach to Vico's thought, «*verum*» et «*factum*» *reciprocantur* ('*verum* [the true] and *factum* [what is made] are interchangeable or, in the language of the Schools, convertible terms') was the basis on which Vico proclaimed that it was human history and not the sciences of which we could hope to have complete comprehension.[1] *Verum* (truth), *factum* (all human artefacts – law, marriage and society, for example, but not religion, according to Vico), and *certum* (certainty; knowledge which comes only from creating something) were the three key terms in this argument. That *verum* was convertible with *factum* indicated that human creations could be accepted as both truthful and legitimate. No such guarantee or relationship could ever be hoped for in the study of nature, which God created and thus only He comprehended fully. According to Vico 'because man is neither nothing nor everything, he perceives neither nothing nor the infinite' ('*Homo quia neque nihil est, neque omnia, nec nihil percipit, nec infinitum*').[2]

In 1711–12 Vico wrote in response to an attack on *De antiquissima italorum sapientia* in the *Giornale dei letterati d'Italia* (Venice):

– Fa' del proposto teorema una dimostrazione – , che tanto è a dire	Create the truth that you wish to [analyse]; and I, in

quanto: – Fa' vero ciò che tu vuoi conoscere – ; ed io, in conoscere il vero che mi avete proposto il farò, talché non mi resta in conto alcuno da dubbitarne, perché io stesso l'ho fatto.	[analysing] the truth that you have proposed to me, will [make] it in such a way that there will be no possibility of my doubting it, since I am the very one who has produced it.[3]

Vico seldom used the terms *verum* and *factum* (or, indeed, the Italian equivalents, *il vero* and *il fatto*) in *La Scienza Nuova*. More often he discussed knowing (*scire*) and making (*fare*):

Lo che, a chiunque vi rifletta, dee recar maraviglia come tutti i filosofi seriosamente si studiarono di conseguire la scienza di questo mondo naturale, del quale, perché Iddio egli il fece, esso solo ne ha la scienza; e traccurarono di meditare su questo mondo delle nazioni, o sia mondo civile, del quale, perché, l'avevano fatto gli uomini, ne potevano conseguire la la scienza gli uomini.	Whoever reflects on this cannot but marvel that the philosophers should have bent all their energies to the study of the world of nature, which, since God made it, He alone knows; and that they should have neglected the study of the nations, or civil world, which, since men had made it, men could come to know.[4]

Vico's conviction that God created man is not inconsistent with his view that history is the one area which offers complete comprehension because it was man-made, *factum*. Vico's argument has nothing to do with the creation or reproduction of mankind. It was the social institutions created by man, the 'world of nations', which was his concern. For in this sense, man made the 'world of nations' (*il mondo delle nazioni*): society and government.[5] Vico's concept of the 'world of nations' is part of a long debate concerning *jus gentium* (law of the nations) and *jus naturale* (natural law). In the seventeenth century, thinkers argued the Maker's Knowledge Tradition, as it is called by Antonio Pérez-Ramos.[6] For Thomas Hobbes (1588–1679) it was the state which is man-made and thus comprehensible, for Robert Boyle (1627–91) and John Locke the emphasis was on knowledge of the physical world. In terms of political philosophy, Vico is more Hobbesian than Lockean (although Vico repudiated the social contract in any form), for he considered society and government began together. For Vico, society was the creation of generations of slow development, whereas for Thomas

Hobbes (1588–1679) society and government began together at the time of the social contract, and for John Locke society began in the state of nature, but government at the time of the social contract.

According to Vico it was necessary to isolate the principles of this *scienza nuova* that are in the human mind and at the same time initiate human society. The prime ingredient was the desire for expression, which was not a privilege or a hobby in early societies, but a requirement. This need for answers, this curiosity, was responsible for the creation of social institutions. Ultimately both the creation of *il mondo delle nazioni* – civil governments and societies – and historical consciousness were aspects of the creative faculty of man – *fantasia*. The truth (*verum*) which Vico sought was knowledge concerning past, man-made societies.

There is a strong connection between Vico's theory of *verum et factum* and Hobbes on the connection between knowing and making.[7] Hobbes wrote that politics is a science and that its truths are demonstrable because men make the commonwealth themselves.[8] Hobbes declared he had founded civil philosophy, the science of political bodies (*De cive*, 1642). In this sense Hobbes considered politics to be analogous to other man-made forms of knowledge. Locke in *An Essay Concerning Human Understanding* (1690) stressed the critical importance of the relationship between making and knowing.[9] But neither Hobbes nor Locke applied the theory of making and knowing to history, primarily because they did not have Vico's fully rounded conception of human knowledge. Hobbes regarded history as an inferior form of knowledge founded on sense, memory and testimony with all their fallibility. Both Hobbes and Locke assumed that for history to be of human making it would have to be comprised of names and definitions given to it by man.[10] In contrast Vico's view of history was not at all dependent on language; rather language was for Vico another human creation. Although his claim is open to debate regarding politics, according to Vico, history was not constructed in the same way as geometry or even politics (deliberately fashioned, in the same manner as making a building); rather he argued history was made slowly and unself-consciously by mankind, as composed of particular social groups. It is at this essential point that Vico's view of history (the civil world, itself *factum*) diverges sharply from Hobbesian commonwealth and Lockean morality.[11]

Vico's concept of *verum* and *factum* had a number of dramatic implications. (1) It called for a radical break between the study of and thus the perception of what exactly were the arts and sciences. This statement in tandem with his later historical principles allowed an attack on scientific and metaphysical Cartesianism.[12] (2) It was because of his theory of *verum* and *factum* that Vico announced that history was the area over which we could gain complete comprehension. (3) *Verum ipsum factum* recognised that human artefacts had characteristics in common in whatever society they appeared, varying similarly age by age, and also that these human creations were the means by which these same societies demonstrated their individual character – by means of their myths, laws and social institutions. (4) This strong statement also implied that these same human institutions could be trusted as true representations of the societies by which they were created, because they were not designed to deceive or magnify (as standard national, political histories inevitably are), but were produced to fulfil actual needs – as explanations of their place in nature (myths) or as a means to live in relative social harmony (for example, accepted marriage patterns and social mores which later became codified laws). *Verum* and *factum* demonstrated Vico's commitment to man-made language and society and the subservience of ideas to institutions: for Vico ideas were generated by early society rather than by rationality (even in a Lockean sense) or by Platonic forms slowly grasped by developing social groups. He did not accept a dualistic epistemology – one for the sciences and the other for the social sciences and history – but rather argued for a single epistemology, which was most fully expressed in the study of past societies.[13]

3. Categories of Historical Knowledge

Isaiah Berlin identified four categories of knowledge discussed by Vico.[1] (1) It was *scienza* (knowledge) which yielded *verum* (truth *a priori*), as created by God. 'Verum presides over what men make – rules, norms, standards, including those which shape the facts themselves; products are known before they are made by the creator' (Berlin).[2] (2) Vico discussed *coscienza* (knowledge of matters of fact, consciousness) as the *certum* that one has of the external world.[3] Berlin's defence of Vico's naming of his best known work is not entirely convincing. Berlin argued that the reason why Vico chose the term *scienza* rather than *coscienza* 'lies in the interplay of what Vico symbolized as the "Platonic" and the "Tacitean" – the general and the particular, the eternal and the temporal, the necessary and the contingent, the ideal and the actual'.[4] The foregoing is no doubt true, but it was not *a priori* truths which Vico sought to discover in history or in his method. It was exactly those specific incidents – human artefacts, past ways of thinking and feeling – that he sought to elucidate. For this reason a much more appropriate name for Vico's last work would have been *La coscienza nuova*, the new historical consciousness of the evolving of social institutions and creations. It should also be kept in mind that a much better English translation of the title of this work would be *The New Consciousness* rather than *The New Science*.

(3) This is not to deny the Platonic patterns, the eternal truths and principles in his writings, notably *'la storia ideale eterna'* (ideal, eternal history; *a priori* knowledge, a pattern to predict and re-predict), which was presented by Vico as an eternal truth. According to Leon Pompa historical laws can be established only if they can be shown to be part of the constitution of historical facts themselves.[5] Since Vico's 'ideal eternal history' was indeed demonstrated by historical knowledge of identifiable patterns in past society, it thus seems unnecessary for it to have been an eternal truth as well.[6] Vico broke with the Platonic tradition, thereby leaving no permanent values or standards. It seems he felt it not simply necessary but stimulating to make his own historical approach an eternal truth itself: doing so was his discovery, his moment of illumination. Now this declaration seems rather pointless, but not surprising in the context of his intellectual background.

(4) The final category of knowledge was 'inner' or 'historical' knowledge, which Vico discussed as knowledge *per caussas* (of causes – Vico's spelling); it was to be attained by attending to the 'modifications of our same human mind' (*'modificazioni della nostra medesima mente umana'*).[7] Knowledge *per caussas* was for Vico the identification of previous events in regular conjunction, of which the causes were generative. It was that which pushed human creation into existence, the dynamic, metaphysical principle, which was connected with very particular views about religion, especially divine providence, that which gives a pattern to history, as well as history itself. Yet it is Vico's method of historical reconstruction, which was neither Stoic nor Epicurean, and the identification of historical sense which we prize today. Ironically the societies he mentioned at most length, particularly classical Rome, tended to be fully flourishing with well-developed civil institutions, especially law, and left particularly rich written records. However it is this final category of knowledge – historical knowledge (very often discussed in connection with the work of Collingwood) – which was Vico's greatest contribution.

4. A Critique of Collingwood

In the English-speaking world it is critical to begin any study of Vico with an analysis of how our view of Vico is coloured by the writings of R. G. Collingwood, philosopher of history, friend and translator of Croce. Vico studies in this century might have developed much sooner (and very differently) outside of Italy if Collingwood had translated Vico himself rather than Croce's monograph on *La filosofia del Giambattista Vico* (*The Philosophy of Giambattista Vico*) in 1913.[1] Unfortunately many of Vico's greatest insights were attributed by Collingwood to Johann Gottfried Herder (1744–1803). The fundamental error was that Collingwood believed Vico accepted human nature as unchanging in all societies at all times, and it was thus to Herder that Collingwood ascribed the concept that each culture must be analysed separately because human nature was constantly changing.[2] Hence Collingwood proclaimed Herder as the father of social anthropology. The most important issue is not which thinker is the accepted head of a particular modern academic discipline, but rather a correct understanding of the concepts involved.

Collingwood was correct in discussing *sensus communis*. However even *sensus communis* itself was not static (as not only Collingwood, but also all other commentators have accepted), and has been adapted slightly in every culture.[3] According to Vico human nature was not unchanging, thus each society needs to be examined individually; but because of common – not preordained – patterns of development, and *sensus communis* (the shared attitudes of all men at all times), which Vico discussed as *il dizionario di voci mentali* (dictionary of mental words), it was possible not only to analyse effectively, but also to make judgements about past societies.

At least three of Collingwood's comments on Vico still require further attention. Arguably the three could be viewed as developments of Vico's thought by Collingwood (and Croce), rather than Vico's own views, but the difference between the two has not been generally apparent to most readers. The first is the 'inside-outside' metaphor. According to Collingwood the outside of an event should be looked *through*, not *at*:

> To the scientist, nature is always and merely a 'phenomenon', not in the sense of being defective in reality, but in the sense of being a spectacle presented to his intelligent observation; whereas the events of history are never mere phenomena, never mere spectacles for contemplation, but things which the historian looks, not at, but through, to discern the thought within them.[4]

Vague as this statement is, it is essential to realize that Collingwood considered that *what* happened in the past was not as important as *why* it happened, because if we can reconstruct why events happened, we can much more easily reconstruct the events themselves. Collingwood has been accused of ascribing 'to historians the role of a psychoanalyst – the capacity to discover in the record of what a man did, various thoughts of which the man himself was unaware'.[5] Patrick Gardiner warned that this approach attributed to historians 'a power of self-certifying insight' (paraphrase by Dray of Gardiner).[6] On these grounds Collingwood – but not Vico – stands accused. Yet, accepting these modern comments as necessary correctives, it is not at all clear why Vico's role of *fantasia* (which was by no means as extreme as that of Collingwood) in the work of historians should be disparaged and not encouraged. In addition it must be noted that Collingwood was more committed to recreating past events than was Vico, who wished to discover the ways of thinking and feeling, the mentality of a particular age or culture.

Second, according to Collingwood, once the historian knows what happened, he already knows why it happened, unlike the natural scientist.[7] For Collingwood events were not events, but an expression of people's characters, what they knew and how they reacted. This Hegelian or Crocean Idealism of Collingwood was actually a good deal more hopeful than Vico's idea of reconstructing the spirit of past ages from past *events*. (Nonetheless it must be restated in Collingwood's defence that he thought it was possible to think *through* the 'outside' of an event to the reasons that caused it).[8] More plausibly, Vico stressed the importance of language and mythology as examined by philology and etymology as the most faithful means by which to reconstruct past societies. On this point Vico appears much more convincing than Collingwood, not only because knowledge of these past events has been lost in most cases, but also because there is no magic key in Vico (or elsewhere) which unlocks the mind of a people from its succession of rulers, battles or other decisive events, even if they are known.

Finally there is Collingwood's most contentious view: in order to understand a human action, the historian must not only discover the thought that expressed it, but must re-think or re-enact the thought in his own mind. This view was expressed most forcefully in the section of *Idea of History* entitled 'History as Re-enactment of Past Experience'.[9] Empathy was the key – if one knows what men were (what it was for them to feel, eat, walk, pray or hate) one will know what they knew. Somehow Collingwood discussed this power of entering into the spirit of another age without relating it to *fantasia* itself. This view is taken directly from Vico, who exhorted his readers *entrare* (to enter) and *descendere* (to descend) into the minds of these *grossi bestioni* (gross beasts) if they ever hoped to have any understanding of these primitive peoples.[10] The crucial difference between Collingwood and Vico on this point was that Collingwood regarded this imaginative method to be infallible. Vico accepted the very real chance that the results might very well be wrong. Vico stressed (and Collingwood ignored) the essential point that having made use of imaginative reconstruction, the historian must then check the results as much as possible by empirical means. This particular view of history being a vicarious experience has been violently attacked in the last half century as a wholly inadequate and subjective means of approaching the past. Although as Louis Mink (1921–83) in his well-known defense of Collingwood's account of historical knowledge wrote '. . . Collingwood's theory of history is so often accepted in principle but disparaged in detail, especially by historians.'[11] What Collingwood's (and thus to some extent Vico's) critics have failed to produce was any alternative method. It is true that philological and etymological research could be done on mythological and linguistic evidence left from past societies without more personal involvement by the researcher. This would be similar to the type of historical reconstruction which Collingwood himself disparagingly described as the 'scissors and paste' approach. If past societies are to be truly resurrected and the spirit of their times identified, it can only be, according to Vico, through the use of *fantasia*, by the imagination and the rational skills of the modern scholar, as applied to a particular period of the past.

5. Historical Cycles

Although Vico used to be hailed primarily as one of the great speculative philosophers of history, along with Hegel, Karl Marx, Oswald Spengler (1880–1936) and Arnold Toynbee (1889–1975), for his three stages of a civilisation, he is generally much more of interest now as an analytical philosopher of history, for his insights regarding historical reconstruction and historical knowledge.[1] Nevertheless the cycles were not an unimportant part of Vico's ideas about history. Only since Marx have some considered it essential to be an economic determinist in order to recognize a pattern of development and decline in societies. It is in this manner that his cycles are so often interpreted.[2] Vico's cycles have often been used by his commentators as a barrier between man and his creations, which both comprise and include history. Vico is perhaps best known for his cyclical philosophy of history, and he is very often connected at the beginning of the chain of thinkers whose best-known exponent is Marx. The Hegelian interpretation dominated Vichian scholarship until after World War II. Although this connection brought Vico some little prominence – notably in nineteenth-century Germany – it had a stultifying effect on any critical analysis of him on his own merits as distinct from a study of him as a harbinger of Hegelian thought. Pietro Piovani's article '*Vico senza Hegel*' ('Vico without Hegel', 1968) was an important turning point in this respect.[3] Vico's cycles are now viewed as a means to examine history through an analysis of the birth and development of human societies and institutions.

The concept of historical cycles was not original with Vico, for cycles can be found in the writings of Plato, Polybius (205?–?125 B.C.), Niccolò Machiavelli (1469–1527) and others.[4] It is not surprising that Vico used this Platonic pattern, when one considers that Plato was the first of his acknowledged *quattro autori*. Vico used the historical cycles, this Platonic pattern, as the foundation for his *storia ideale eterna*, which in turn he considered to be the foundation of his *scienza nuova*.

He maintained that the cycles were a means to examine societies, particularly pre-literate ones. This was due to the special, general character that each of these three periods possessed. For Vico these cycles were not cyclical but spiral-like, and they were also open-ended. He was convinced of the individual character of particular societies, which was not blurred by the features it had in common with any other society at the same stage. This is the *corsi e ricorsi* (course and recourse) of the nations to which he referred. These epochs, which he named the ages of gods, heroes and men, tended to recur in the same order in any and every society until the decline into a new 'barbarism of reflection' (*'la barbarie della riflessione'*), at which point the people are 'rotting in that ultimate civil disease' (*'se i popoli marciscano in quell'ultimo civil malore'*). Only then did he regard it as clear that thought in that particular society had exhausted its creative power.[5]

> First, the guiding principle of history is brute strength; then valiant or heroic justice; then brilliant originality; then constructive reflection; and lastly a kind of spendthrift and wasteful opulence which destroys what has been constructed.[6]

This was Collingwood's elegant description of the Vichian pattern. Although modern Vico scholarship accepts the cycles as a model of historical change in society, not as preordained, many interpreters see the barbarism of reflection inevitably programmed into the final stages of every society, although this pattern does not logically follow from their view of the cycles.

Vico's view of the barbarism of reflection was unusual. His fear of the doubtful, sceptical, cynical outlook was based on the belief that criticism destroys something essential in society. He argued that the existence and development of society was based on the organic links between human needs, based on common beliefs which are taken for granted. The relation of barbarism of reflection to the rest of Vico's thought has not been fully appreciated by scholars, who have tended to neglect the issue of *senso comune*, shared values. Yet a sustained analysis of his work leads to the conclusion that barbarism of reflection occurs when people lose contact with and, indeed, begin to question *senso comune* itself. Whether this catastrophic effect of intellectual development was inevitable for a society, it is clear that Vico believed that intellectual sophistication did to some extent carry with it the seeds of its own destruction, because the creative instinct in that society would have died out by that point.

6. Historical Sense

Vico was not precise about how men make their own history. Very often a parallel is drawn with Marx, who stated that 'man makes his own history, but he does not make it out of the whole cloth'.[1] A more modern translation of Marx renders this passage as the following:

Die Menschen machen ihre eigene Geschichte, aber sie machen sie nicht aus freien Stücken, nicht unter selbstgewählten, sondern unter unmittelbar vorgefundenen, gegebenen und überlieferten Umständen.

Men make their own history, but they do not make it just as they please; they do not make it under circumstances chosen by themselves, but under given circumstances directly encountered and inherited from the past.[2]

It could be argued that these same sentiments were implicit in Vico, but to some extent to do so is to make excuses for Vico's neglect of physical, biological and mental factors, to give only a few examples.[3] It would be more correct to admit that Vico was completely uninterested in almost all these additional elements. History for Vico meant the pattern of development in past cultures and the idea of society. Accordingly, any aspect of the past was historical (in his sense of the word) only if it involved social relations and creations. He designed theories of historical reconstruction in order to make contact with the mentalities of past cultures. The historical knowledge he sought had to do both with ways of approaching these past ages and knowledge of their means of expression and personality.

Berlin has asked what it was that first planted in Vico's mind the awareness of the diversity of cultures. One possible answer may be the biography (virtually a hagiography) of Antonio Caraffa, a Neapolitan general, which Vico wrote for pay. Vico was commissioned by a former student of his, the nephew of Caraffa, Hadrian Caraffa, Duke of Traetto, to write this biographical account. In return Vico received sufficient funds to pay the dowry of his daughter, Luisa, the next year; in addition he gained the friendship of Gian Vincenzo Gravina (1664–1718), a Calabrese writer, at the time much more famous than Vico, who greatly admired the study of

Caraffa. *De rebus gestis Antonj Caraphaei (The Life of Antonio Caraffa)* was published in 1716, but Vico had earlier written a *canzone* (lyric poem) for a Caraffa family celebration in 1693 – six years before he delivered Oration I.[4] Such set pieces were quite common at this time in Italy; a family would commission a local writer to compile a *raccolte* (little collection of poems) on the occasion of a marriage, funeral or taking of the veil.[5] One guess might be hazarded that these quasi-historical pieces which Vico wrote for much-needed commissions might perhaps have contributed to his disgust with contemporary writing and helped lead him into more theoretical speculation.

This was Vico's one experience of using primary sources, Caraffa's own papers, and it was not a happy one. A general in the Austrian service during the Hungarian wars, Antonio Caraffa's name is still remembered with horror in Hungary. However, not surprisingly, Vico's laudatory account ignored the butchery and stressed instead the decadence of the Ottoman empire.[6] The Hungarian revolt was the unifying theme of the work and Caraffa served as a somewhat minor figure in his own biography. (For example, Nicolini translated fourteen pages of extracts from this biography into Italian, only one sentence of which dealt with Caraffa.) Vico, recognizing his need to put these conflicts into an international context, turned to Hugo Grotius, who was to become his fourth *autore*.[7] For some time Vico considered editing an annotated version of Grotius's collected works, but later rejected this as an inappropriate study for a Catholic because of Grotius's deism. Even so his study of Grotius on natural law was to prove more important to his developing theories of history than the Tacitean model he so admired, which had to do with political decision-making in a specific historical context.[8]

The convoluted approach, which is the rule rather than the exception throughout Vico's writings, is readily apparent in *De rebus gestis Antonj Caraphaei*. But happily, after long pages of Vico's digressions, there appears, as so often happens in his work, an excellent and fairly concise section, dealing with the customs of nations, the decline of empires, and divine providence, all three important Vichian themes.[9]

Greatness and decline in the Ottoman Empire fascinated Vico, but there is little of the liberal tone he was to use in his theoretical discussions of the development of different cultures. The Koran was called a lying code which promoted servility and the fatalistic attitude that nothing could change the predestined moment of death. Bribery, trickery and luxury were the recurring themes in Vico's portrait of Ottoman society. He made little mention of women – except for a pedestrian remark that women were manifestly incompetent in civil matters and that in private they dressed well – or polygamy. One would have expected that polygamy in advanced societies would have greatly intrigued Vico, representing as it did a deviation from his set concepts of marriage and family, so fundamental to his theory of the development and decline of nations. Vico's horror of the punishments dispensed in the Ottoman Empire, such as the example of entire families being put to death with the prisoner, could well be termed proto-Beccarian. It was not just the Sultan and Grand Vizer's lack of concern for the human rights of their subjects which amazed Vico, but also the very foreignness of their outlook. This observation was very much in line with Vico's view: that past societies were probably much more dissimilar to one's own society than would be expected. There was in this account some of the ambivalence towards such a diverse culture, although none of the same elegance or originality, that one finds in Montesquieu's (1689–1755) *Les lettres persanes* (*The Persian Letters*, 1721). Nevertheless this (in most respects justly maligned) historical work by Vico reaffirms what is so evident from his theoretical writings, and may well provide at least a partial answer to the question of where he gained the idea of the diversity of cultures.

Indeed in the final edition of *La scienza nuova* there were few references to specific primitive societies (North American Indians, for example) – unlike Montesquieu, Rousseau and Voltaire (1694–1778), in whose works there appeared virtually countless references to contemporary, historical and often fictional primitive societies.[10] Vico was undoubtedly much more concerned with the *idea* of primitive societies than with any particular examples. For Vico history had to do with past cultures and society and with a sense of historical awareness – he was much less interested in the way we write history.

Vico's concept of history was based on two Renaissance ideas: that historical reconstruction was more than a catalogue of past events and that even the advanced civilisations of the past were a good deal less like our own than the natural law theorists had argued.[11] Vico's view of culture was undoubtedly pluralistic, yet he seldom discussed the modern definition of pluralism as various social groups with different native languages, customs, religions or political beliefs living together in relative harmony. The pluralism Vico addressed had to do with a collection of distinct cultures and was based on the belief that there was no ideal state or superior pattern of development by which all other societies should be judged. His discussion of cultures was certainly contemporary as well as historical and he saw no break between the two. Thus each culture had the right to develop as it wished, and it was only as it exercised this prerogative that its true abilities or genius could be manifested.

Vico's 'first indubitable truth' was that 'the world of nations must therefore be found' (*ritruovare*, retrieved, recovered, discovered) in the modifications of our same human mind' (*1744*, §349).[12] Perez Zagorin, who has written one of the very best of the recent articles on Vico, argues that 'there is no common mind available, and that there are far too many variations'.[13] Zagorin is quite right not to accept all of Vico's theories uncritically. Nevertheless, this 'common mind', '*nostra medesima mente umana*' ('our same human mind'), is an essential if problematic part of Vico's thought. Following from Zagorin's assertion that there is no common mind, '*una certa mente umana delle nazioni*', Zagorin states that the historian does not know the past in his generic character, but solely as the particular man he is.[14] But if we were to do away with *mente*, shared consciousness, there would then be no means of making judgements or of analysing past or contemporary lifestyles different from our own. Certainly 'if the past is entirely alien, then it is a past that we will never know or comment upon in any way'.[15] Various definitions for Vico's *mens, mente* have been suggested: universal mind, Hegelian *Geist*, Jungian 'collective unconsciousness', or *sensus communis*.[16] It is not at all clear that *mente* was for Vico, as has been argued, the creative principle of the world – this role was explicitly reserved by Vico for *fantasia* – nor that ultimately *mente* was the same as the concept of culture (again, this is *fantasia*, as the spirit of particular society, not the totality of the individual society itself).[17] At the very least it can be said that Vico's *mente* was intimately related to *sensus communis* and his *dizionario di voci mentale* and thus its importance cannot be doubted. *Mente* has to do with what is common to and alive in society, such as social relationships. It is that which generates culture (although not culture itself), an active faculty, not thought but a non-Jungian, collective human consciousness which develops and perhaps even understands itself.

Although Vico based his historical theories on *verum ipsum factum* it was not his intention to discover the truth in history. He desired to reclaim human artefacts, social institutions. Vico considered myths to be true in the sense that they were representative of the time in which they were created. He believed myths were not false statements about reality. At the same time they were not true in any abstract, metaphysical manner. Neither did Vico seek to reclaim particular facts or events. Chronological reversal – reading history backwards – was one of Vico's approaches to history, but it was not the chronologies themselves which were important to him.[18] Berlin wrote, in a particularly Vichian manner, that 'facts are not hard pellets of experience, independent of concepts and categories by which they are discriminated, classified, perceived, interpreted, and indeed shaped'.[19] All history for Vico was a history of ideas; he was much more concerned with explanation than truth or facts in history. Since a reconstruction of past ideas and ways of thinking can never be complete, in the same way 'there can be no complete

or finished or definitive history . . . only partial, incomplete, even tentative' (Preston King).[20]
   Vico never seems to have agonized over how one can know the past. A modern concern – the desire not to blur the contrast between 'then' and 'now' – was articulated by Vico to the extent that he stressed that people of the past were probably less like us than would generally be thought, but he did not see any inconsistency regarding thinking in the present about the past.[21] He was well aware that any thought we are now thinking is a present thought, and he stressed the need for imaginative reconstruction. That the present contains the past within it was a constant theme in his work, in which he explains why he considered language to be perhaps the best link with the past because it is shaped by the past, is a product of some process of the past and carries with it the marks of the past.[22] Language has to do with both past and present things, then and now. Vico propounded the view that language can only be fully understood in reference to the experiences of the past; modern speech evokes a dim awareness of the origins of the words used, often for an entirely different purpose. Vico stated that if his warnings regarding historical anachronisms were heeded, effective analysis of the past could be done by historians with no personal connection to the time they were studying. The role of historians was thus to identify and examine these past cultures, making use of both their reason and their *fantasia*.

7. History as a Science

One of Vico's most useful insights was the identification of the lack of a proper historical awareness, an historical sense. Vico was deeply indebted to neo-Platonic writers, notably Lorenzo Valla, for this approach. The relationship between legal humanism and the study of classical texts and the development of modern historical studies is identified by Donald Kelley in his book, *The Foundations of Modern Historical Scholarship*.[1]
   Vico isolated five major problems in this regard, which are very often viewed as a parallel to Bacon's idols of the mind, as 'fallacies that systematically distort thinking' (Zagorin).[2] First, he discussed exaggerated opinions about antiquity, particularly concerning the national history of each society.[3] Scholarship, battles, establishment of kingdoms and the behaviour of the young, for example, are often spoken of as having been so much better in the past. This longing for happier days of the past is still very much present today and the desire to embellish and aggrandise past events and achievements was seen by Vico as a barrier to any possible realistic understanding and analysis of the past. Second, he discussed 'the conceit

of nations' (*'la boria delle nazioni'*).[4] The feeling that the development of one's own country is of the utmost concern to all countries, the belief that the splendour or dominance of one's country must have been apparent at every stage of its development, and the assumption that one's country is splendid, dominant and best, at least in the areas one considers important – military endeavour, culture, or lifestyle – all of these were attitudes which Vico considered must be acknowledged, at least by historians, even if they could not be eradicated.

Third, Vico discussed the 'conceit of scholars' (*'la boria de' dotti'*).[5] This was a favourite topic of his. According to Vico, scholars tended to think of people in the past as people like themselves, of an academic, reflective outlook.[6] He blamed scholars repeatedly for stifling the imagination of the young. Vico seems to have taken particular pleasure in stating that the most effective men in history were not academics. He called for a new Augustus 'to arise and establish himself as a monarch and, by force of arms, take in hand all the institutions and all the laws, which, though sprung from liberty, no longer avail to regulate and hold it within bounds' (*'E, come Augusto, vi surge e vi si stabilisca monarca, il quale, poiché tutti gli ordini e tutte le leggi ritruovate per la libertá punto non piú valsero a regolarla e tenerlavi dentro in freno'*).[7]

Fourth, he discussed what Collingwood termed the fallacy of sources.[8] It was generally considered that societies must share sources in order to have the same characteristics. Thus one society would have to have borrowed a concept from another or both from a third, if an identical pattern could be identified. Vico made no attempt to deal with cultural sharing. He declared that every society went through a similar pattern of growth, and its stage of development could be identified by comparing it to other societies. The fortunes of Vico's own theories exemplify his view that ideas are not diffused, but created by each society when needed. For example, his exact idea, the identity of Homer, is often cited as a discovery of nineteenth-century Germany.[9]

Finally Vico considered it necessary to remind his readers that societies in antiquity were most probably not better informed than we ourselves about societies that lay closer to them. This statement usually is lost in any study of Vico. However it is a powerful reminder that Vico was setting out not only ways of approaching the past – methods of historical reconstruction – but also delineating the proper way of thinking about the past.

Vico's historical method was constructive as well as critical; one reads Vico not only for the problems he identified but also for the solutions he presented. Vico declared that linguistic, etymological and philological study could shed light on history because it was language which created minds, and not minds language. Vico argued that the poets did not merely create artificial worlds but that mythologies expressed the social structure at the time of their creation.[10] This theory was of particular use to Vico since he maintained that other minds at the same stage of development tend to create the same products. He wrote:

Le tradizioni volgari devon avere avuto pubblici motivi di vero, onde nacquero e si conservarono da intieri popoli per lunghi spazi di tempi.	Vulgar traditions must have had public grounds of truth, by virtue of which they came into being and were preserved by entire peoples over long periods of time.[11]
. . . vi si vaglia dal falso il vero in tutto ciò che per lungo tratto di secoli ce ne hanno custodito le volgari tradizioni, le quali, perocché sonosi per sí lunga etá e da intieri popoli custodite, per una degnitá sopraposta debbon avere avuto un pubblico fondamento di vero.	Truth is sifted from falsehood in everything that has been preserved for us through long centuries by those vulgar traditions which, since they have been preserved for so long a time and by entire peoples, must have had a public ground of truth.[12]

It was this 'public ground of truth' ('*un pubblico fondamento di vero*') which Vico desired to obtain. His was an altogether new manner of thinking about tradition. Although not literally true, these alternative sources were also not false statements about reality. These myths, laws and other social institutions were the primary source materials needed to reconstruct past societies – they provided the most pertinent type of historical information that one could wish. One does not sense that Vico regrets the lack of written records for pre-literate or other societies, rather he positively delighted in this non-literate means of discovering and examining the cultures and mentalities of the past.

2 Vico's Early Writings, 1709–28

1. Introduction

Most writers on Vico have concentrated their efforts on his important final work, *La scienza nuova*, to the exclusion of his earlier writings; yet, within these works – which include an autobiography, orations, an unfinished book on the wisdom of the ancient Italians, a book on law, and the first two editions of *La scienza nuova* itself – Vico addressed the fundamental issues of imagination and history. Many crucial insights regarding these and other issues raised there do not reappear in his later works. Consequently a fully-rounded understanding of Vico's thought is not possible without consulting the 1699–1730 works. The autobiography gives the most detailed description we have of Vico's life and intellectual development. *De nostri studiorum ratione* and *De antiquissima sapientia italorum* are much undervalued as sources of Vico's growing preoccupation with imagination in particular, and it is in *Il diritto universale* and the first two editions of *La scienza nuova* that Vico expressed his ideas on social groups and human history in the most sophisticated form. Thus this chapter is designed to introduce these important works, which will form an integral part of the analysis of Vico's thought in later chapters.

2. *De nostri temporis studiorum ratione* (1709)

As with his earlier orations of 1699–1707, Vico clearly felt no compunction to limit this essentially pedagogical piece simply to the standard debate regarding the relative merits of the Ancients and Moderns – although he was far from indifferent on this point.[1] As in his other orations, he began and ended this one with issues and platitudes of the time, but in between he gave his (no doubt uncomprehending and bewildered) audience a distillation of his own ideas. He discussed standard themes of the time in order to put across his own ideas. He began *De nostri temporis studiorum ratione* (*On the Study Methods of Our Time*) by stating that every study method was composed of three things: instruments, of which the philosophical method was the most

41

common in his time, complementary aids, such as literature, fine arts and printing, and the aim envisaged.[2] This tedious opening notwithstanding and although Vico's stated aim in this work was to compare the ancient and modern study methods, what he wrote was a startlingly clear denunciation of any dogmatic application of Cartesian principles. Twenty-three years later in his final oration, *De mente heroica* (*On the Heroic Mind*, 1732) Vico returned to the issue of the Ancients versus the Moderns and reiterated his call for a balanced approach to study in the university.[3] Unfortunately the final oration had little new to add to his views on imagination and the historical method. The normal university lectures Vico delivered to his students as part of his duties as a professor of rhetoric, now entitled *Institutiones oratoriae* (*University Lectures*) stressed invention, testimony and metaphor.[4] One can find a clear link between his university orations and his own personal writings, the sharp divide between the two, so often stressed by Vico scholars, robs one of additional Vichian texts.

This emphasis on the proper approach to academic study was first stated in *De Nostri temporis studiorum ratione* and, not incidentally, he combined this with a forceful statement concerning the need for a new historical method:

Ac haec addas librariorum menda,	We must guard against scribal
librorum plagia et imposturas,	garblings, plagiarisms, forgeries,
alienae manus irreptiones, quibus	interpoloations of alien hands
legitimos authorum partus vix	through which it is difficult for
agnoscimus, vix germanos sensus	us to recognise the originals, and
assequimur. Ita ut, cum,	to grasp the author's true
quod nos scire oportet, tot	meanings. What we need to know is
libris contineatur, quorum	contained in so many books in
linguae intermortuae, respublicae	languages that are extinct,
deletae, mores ignorati, codices	composed by authors belonging to
corrupti, una quaevis ars	nations long since vanished. These
scientiaque adeo difficilis	books contain allusions to custom
facta est, ut vix singuli ad	often unknown, in corrupted
singulas profitendas	codices; therefore the attainment
sufficiant. Itaque studioruom	of any science or art has become so
universates nobis institutae	difficult for us, that at the
sunt, et omni disciplinarum	present time no person can master
genere instructae, in quibis	even a single subject. This has
alii alias doctrinas, suae	made the establishment of
quisque scientissimus, tradunt.	universities necessary.[5]

As was his way, Vico used this occasion primarily as a platform for discussion of many of the themes which played such a vital role in all his theoretical works – including *sensus communis, phantasia, ingenium* and, to a lesser extent, memory, imitation and poetry. Vico's writings in general represent a shift in emphasis to the issues of society, culture and history, away from the usual stress on politics and government. This work is no exception.

On 15 November 1708 when Vico gave this oration as the annual inaugural address at the University of Naples, he once again drew the analogy of the young as strong in *phantasia*, the old in reason. He stressed common sense and memory, because it was the faculty closest to imagination in the later stages of a society. He urged that *phantasia* should be cultivated in the young.[6] For this reason, he argued, subjects such as geometry, which demanded both memory and imagination, were to be strongly supported. Vico assigned to memory, as well as imagination, very much more than a passive function.

Vico mixed his discussion of oratory and other subjects at the university with his own views concerning the study of culture and society. For example in the midst of a discussion on medicine and the natural sciences, he made perhaps his only remark on ethics.[7] Never a great interest of his, perhaps as it had as much to do with individuals as social groups, ethics received relatively little attention in this period, because of the increased interest in the natural sciences. Vico's mention of ethics here is explained when it is realised that it was a mixture of demography and sociology to which he was referring, rather than a set of standards which individuals or even groups should strive to achieve. He feared that the increasing dominance of the sciences would lead to the neglect of ethics, politics and 'human character, dispositions, passions and the [correct] manner of adjusting these factors to public life and eloquence' ('*quae de humani animi ingenio eiusque passionibus ad vitam civilem et ad eloquentiamaccomodate*').[8]

He expressed a desire for a broadening of the university curriculum, so as to allow for more study of other past cultures. It was not his intention to promote narrow specialisation, but rather to encourage a more balanced approach in which students would be taught both arts and sciences, and thus more about society and history, particularly concerning the ways in which previous generations and cultures lived their lives. Vico stated that universities failed to enquire into human life and that consequently the students were not prepared for the actual life they would lead after their education was completed.

Throughout this work Vico stressed the importance of the practical aspect – sometimes as *sensus communis*, more often as practical judgement [*prudentia*].

Nam prudentia in humanis actionibus vestigat verum uti est, etiam ab iprudentia, ignorantia, libidine, necessitate, fortune: poësis tantum ad id verum spectat, uti natura et consilio esse debet.	Practical judgement in human affairs seeks out the truth as it is, although truth may be deeply hidden under imprudence, whim, fatality, or chance; whereas poetry focuses her gaze on truth as it ought be by nature and reason.[9]

As vague as this statement is, the implication is clear enough that Vico felt that modern, more sophisticated modes of analysis overlooked the simple and the obvious.[10] One of his greatest fears, in regard to philosophical criticism, was that it would swamp the *sensus communis* of the young. Hence he considered it necessary for students to spend some part of their youth uncontaminated by the sophistry of both modern and ancient thought, during which time their intuitive wisdom, shared with all members of their society, their *sensus communis*, would develop sufficiently to withstand the onslaught of their further academic training.[11]

Vico feared that subjects which depended on *sensus communis* would in time be systematised into inactivity. He complimented geometry by stating that it encouraged ingeniousness.[12] This was particularly important to Vico and he often stated that only ingenious minds could produce new inventions. He noted with pleasure the usefulness of printing, which allowed new, untried authors to be published much more easily – a great personal interest of his – and discussed in some detail the disciplines which depended on sound or practical judgement: oratory, poetics and the art of writing history.[13]

In the midst of a brief discussion concerning the problems of blending the Cartesian geometric method and Aristotelian physics, Vico offered one of his innumerable lists of three: abandon the new method altogether, incorporate it within the old method or retain the older method used at present but account for any new phenomenon as a corollary to this modern type of physics.[14] In spite of these half-hearted attempts to mix the old with the new, Vico could not be reconciled to the widely held belief that Cartesian geometric physics was the authentic voice of nature. This popular view was diametrically opposed to that which he was to express most explicitly in his next work, *De antiquissima sapientia italorum*. Vico asserted that any archetypal forms, ideal patterns of reality, existed in God alone and could never be fully comprehended by man ('*In uno enim Deo Opt. Max. sunt verae rerum formae, quibus earumdem est conformata natura*').[15] Thus the pursuit of science was for him less compelling, simply because it could never be grasped fully by man, whereas the study of history and of society offered the best opportunity for rigorous and complete study.

One might have expected Vico, who had borrowed so liberally from Bacon in so many other areas, also to have embraced the experimental method, which had been the focal point of the scientific academy, *I Investiganti* (The Investigators), which dominated Neapolitan intellectual life in the previous generation. The experimental method would seem to be particularly relevant in the midst of this particular oration, in which he repeatedly stressed the importance of practical knowledge. But, as is abundantly clear from his later works, Vico was always inclined to the theoretical rather than the pragmatic approach; on those occasions when he did discuss the practical aspects of an issue he was almost invariably dealing with early stages of development, not sophisticated technical advances. Most notably, Vico was concerned at this point to develop a new concept of wisdom. He considered abstract knowledge to be the highest form of truth, *sensus communis* the lowest.[16] The important point is not the ranking of *sensus communis* as the lowest form; rather the issue is that Vico considered it to be a form of knowledge at all – indeed for Vico it was the basis of all knowledge. Even in the advanced stages of a civilization, Vico maintained that abstract knowledge alone was not sufficient, for *sensus communis* was necessary if such advanced notions were to be communicated to a larger audience in order to save developed societies from falling into the traps of luxury and laziness. Vico maintained it was 'impossible to assess human affairs by the inflexible standard of the abstract right' ('*non ex ista recta mentis regula, quae rigida est, hominum facta aestimari possunt*').[17] There are are modern parallels of Vico's *sensus communis* in the work of the philosopher Donald Davidson on the possibility of translation and of comprehending other cultures.[18]

Chance and choice played the dominant roles in human affairs, according to Vico. Thus any educational system and particularly any study of history should reflect the variety of human actions and intentions, the ambivalences of life and fluctuations in fortune of all sorts. Vico noted that even the art of writing history varied directly with time and place. This view clearly reflects Vico's notions of the changeability of human nature and the freedom of man's will. An example of practical wisdom was for him a view of an event which would accord the greatest number of causes to it.

Vico considered *sensus communis* to be not only the criterion of practical judgement but also the guiding standard of eloquence. Contrary to the prevailing Cartesian view, eloquence, according to Vico, was not to be downgraded. Indeed this art of speaking the truth in public was to be actively encouraged. It can safely be assumed that Vico regarded the art of eloquence as one of the principal means by which the young should be taught. He wrote in '*La pratica*', the final section of *La scienza nuova seconda* unpublished in his lifetime, that if this civic training was executed properly, the cataclysm of his third and final stage of a society could be averted, or at least postponed indefinitely.

The theme of imitation tantalizes the reader of this work, for there is in *De nostri temporis studiorum ratione* the well-known phrase that 'without Homer there would have been no Vergil, and without Vergil no Tasso' ('*Neque–enim, inquiunt, esset Virgilius, nisi ante fuisset Homerus: neque apud nostros Torquatus, nisi ante Virgilius*').[19] Thus Vico implied the continuity of the great themes, but from the 1744 work it can be seen that he also argued very strongly against the traditional belief that there had been a single person who had written all the works attributed to a man called Homer. The thesis of '*Della discoverta del vero Omero*' ('On the Discovery of the True Homer' (Book III of the second and third versions of *La scienza nuova*) was that the Homeric epics were the distillation of the ancient wisdom of the Greek people, handed down generation by generation and eventually written down. On the one hand, Vico did not despise imitation, rather he thought it to be the natural way in which a society passed on its own history. In the same vein he stated that one's reading should be governed by the judgement of the ages. Vico's ideas were drawn from Renaissance concepts of origin and originality. According to David Quint the tension between tradition and modernity in Renaissance literature was due both to the desire to identify classical sources for modern ideas in a society becoming increasingly aware of itself historically and to the new appreciation of contemporary literature and art.[20] On the other hand, Vico clearly stated in *De nostri temporis studiorum ratione* that a genius does not model his work on established masterpieces, for it is not possible to produce anything of lasting merit if it is simply copied from what has gone before; he even went so far as to write that the most outstanding masterpieces hinder rather than help students in the field. In order to produce something original a break must be made eventually from the old masters and the inspiration must come from nature, by which he meant in this case both the social and physical environments.

The richest passages in *De nostri temporis studiorum ratione* were those in which he contrasted the concepts of *phantasia* and reason, poetry and metaphysics, and the poetic and logical faculties. Through these comparisons the clearest insight can be gained regarding his use of the individual terms. Poetry, for example, was mentioned only occasionally in this work, once described as a gift from heaven; at another point, he stated that there was no instrument, no artificial means, by which poetry could be attained, but he did not refine these ideas beyond this point.[21] With regard to the concepts of language and poetry, *De nostri temporis studiorum ratione* offers little new to anyone well versed in *La scienza nuova*, but in terms of *phantasia* and *sensus communis* in particular, the key elements in Vico's ideas regarding history and society, this oration is fundamental.

Without a doubt, the single most important statement Vico made in *De nostri temporis studiorum ratione* was to condemn the French language as unsuitable for either stately prose or for verse. At the same time he lashed out at the French intellectuals for praising the kind of eloquence which characterised their language.[22] This outburst did not simply reflect Vico's linguistic xenophobia, but went much farther, for it was meant as a direct criticism of the Cartesian approach, which by this time had a strong hold not only in northern Europe, but in Naples as well. Unknowingly, Vico was much more in agreement with British than French thought of his time. There was an eighteenth-century move away from rationality among many British writers, including Anthony Ashley Cooper, Earl of Shaftesbury (1621–83), David Hume (1711–76), Smith and Edmund Burke (1729–97).[23]

Vico objected to study methods, and even languages, which by their very construction defined which subjects should receive attention and even how such restricted areas should be pursued. Vico desired to throw open the fields of enquiry, and he viewed Scholasticism and Cartesianism as strange bedfellows in the ruling academic establishment, equally determined to thwart his new approach. Thus as early as 1708, thirty-six years before the final version of *La scienza nuova* was published, Vico had already drawn up the lines of his attack on the rationalist emphasis of his time.

3.   *De antiquissima italorum sapientia* (1710)

Vico's readers might have expected this work to include a familiar discussion of Aristotle and the scholastic tradition, the Ionians, the Etruscans and ancient Italian thought. They were not to be disappointed. Just over fifty pages long, much of *De antiquissima italorum sapientia* (*On the Wisdom of the Ancient Italians*), as with *De nostri studiorum ratione*, would be of more interest to historians examining the institutional history of the university – for example, a study of the curricula – than those interested solely in the development of Vico's thought. The title was most probably borrowed from Bacon's *De sapientia veterum* (*On the Wisdom of the Ancients*, 1609).[1] Vico's book was to have been in three parts: metaphysics, physics and morals, but in the end he wrote only the first part on metaphysics. Nevertheless it was in *De antiquissima italorum sapientia* that Vico made his clearest and most forthright statements regarding *verum* and *factum*: that man could never fully know the natural world, and hence master the sciences, because the natural world was made by God, but that human history was largely if not entirely comprehensible precisely because it was man-made.[2] For this reason *De antiquissima italorum sapientia* has long been taken to mark a major epistemological break in Vico's thought. Despite that, as was discussed above, these themes were present in the orations and *De nostri temporis studiorum ratione* as well.

Vico made constant references throughout this work to a host of ancient Italian thinkers. Sometimes it was only to raise the question as to whether they agreed with Aristotle, the Roman legal writers or some other well-known authority. But the title of the book is a sham, because nowhere in the work does he actually discuss whether there was a distinct school of thought which could properly be termed the ancient wisdom of the Italians, nor did he specify to which (if any) of the ancient Italic and post-Roman peoples he was referring. There is no discussion of any indigenous developments, only these passing references to outside influences. There is indeed no evidence whatsoever that such ancient Italian thinkers ever existed; these fictional characters seem to have been invented by Vico to lend authority to his own original ideas regarding *verum* and *factum*, for which he had no classical or modern sources.[3] Indeed even a reviewer of the book at the time stated that there was no proof at all concerning these ancient wise Italians. (Otherwise the review was quite positive, even though, or possibly because, the reviewer did not grasp the originality of Vico's argument. Perhaps for this reason Vico took offence at some rather trivial points and wrote two responses, which were also published in 1711 in the *Giornale de'letterati d'Italia*.) Nevertheless Vico's references to ancient Italian peoples were later taken up by Italian nationalists eager to find a discussion of pre-Roman, Italic peoples.[4] Yet there are no traces of proto-nationalism in Vico's writings, in the vein of the final chapter of Machiavelli's *Il principe* (*The Prince*, 1513) and certainly no condemnation of classical Rome.

The established Latin writers to whom Vico referred so frequently argued that everything produced by the mind was entirely the result of sense perception. For Vico this was clearly an example of pagan metaphysics, because Christian metaphysics taught just the opposite. He did speculate as to whether the ancient fictional Italian philosophers accepted, with Aristotle, that the human mind perceived nothing but by the senses, but he seemed not to have been particularly interested in the answer.[5]

Of much greater interest than his discussion in *De antiquissima italorum sapientia* of the five physical senses and their relationship to the mind was Vico's emphasis, once again, on *sensus communis*.[6] Vico scholars have overlooked the issue that he did not believe that every group of people shared the same type of *sensus communis*, as to have done so would have presupposed an identical pattern of development. Rather he maintained that all peoples have their own version of *sensus communis*. In this way his approach was more complex than has previously been recognised. For *sensus communis* was a faculty shared by all social groups, but which was manifested in a multitude of forms. When Vico examined in *De antiquissima italorum sapientia* what common sense is (*'Quid sit sensus communis'*),[7] his answer was 'the likeness of customs among peoples gives birth to common sense' (*'Similitudo autem morum in nationibus sensum communem gignit'*). This issue was addressed in the properly rhetorical sense, not that it was so obvious that it need not be asked, but rather that it was a teaching device to illustrate a point he considered to be most important to stress. This use of rhetoric was to be found in many of Vico's authors, Bacon being the best modern example.

One of the major themes of this book was the relationship between *memoria* and *phantasia* which Vico considered to have been almost symbiotic, but he made the issue less straightforward by also stating that 'men can remember nothing not given in nature' (*'Homini fingere nihil praeter naturam datur'*).[8] It was not at all clear whether he meant in this case nature as the environmental conditioning of a child by the physical and social world into which it was born, or if it represented the native abilities of each child. Either interpretation diminished the importance of imagination for the individual, once again stressing his emphasis on social groups, not individuals, as the important factor in the development of the human mind. Likewise memory was not that of a particular person, of events glimpsed, and attitudes sensed or even misunderstood during youth; rather, it was the common memory shared by all people of a particular social group, hence memory and imagination formed in effect a collective *phantasia*.

Phantasia was not at all a subsidiary attribute for Vico, nor was it to be considered simply as a creditable gift or talent of a people. For Vico imagination was an essential, true faculty because 'we are creating images of things' (*'Phantasia certissima facultas est, quia dum ea utimur rerum imagines fingimus'*) and this idea led to the well-known statement that 'it is when we understand something that we make it true' (*'Ad haec exempla intellectus verus facultas est, quo, cum quid intelligimus, id verum facimus'*).[9] Images played a pivotal role in Vico's design, for his whole orientation was towards the discernment and interpretation of past attitudes rather than of specific events. Thus in Vichian terminology images (perceptions and memories) were true – not because they were necessarily accurate historically, but because they were in and of themselves authentic representations of the essence of former times. Thus Vico's own work on

Caraffa was true, in the sense that it was a piece of primary evidence of an account written by a younger contemporary at the time of the general's death, even though it was wildly inaccurate historically.

Vico himself advised caution when meditating on truth. For, he wrote, the line between passions and prejudice was very often difficult to recognise, even when it was consciously sought. This statement has a deeper significance in the context of his analyses of attitudes towards the past. If prejudices and biases, which are both natural parts of past outlooks, are not acknowledged, if only by the onlooker of himself, then the result can simply be another piece of primary evidence – for example, how an eighteenth-century Neapolitan academic viewed previous civilisations – rather than an attempt at theoretical historical analysis, which would, of course, be an example of secondary evidence. Needless to say, Vico somehow considered that he could write about bias without exhibiting it.

Elsewhere in the book Vico defined human knowledge as the dissection of the works of nature, thus inferring that all knowledge was simply the gradual comprehension of nature itself, insofar as that was possible, the most important aspect being the increase of the mental abilities of human beings.[10] Yet the physical world, which at other points he described as nature, was, as is well known, excluded from his method of historical understanding. Arithmetic, geometry and mechanics, he wrote, lie within human faculties, but physics he regarded as within the faculty of God. Nevertheless it was from nature that he drew many of his analogies, as when he also compared human knowledge to chemistry, by which one assumes he meant that there was a systematic body of knowledge that was necessary for a proper understanding of the human mind, just as there was in a science such as chemistry, based on the acquisition of a standard body of work. Hence although *De antiquissima italorum sapientia* is quite rightly regarded as anti-Cartesian, it is not anti-scientific. Vico happily used examples from mathematics and the natural sciences to back up his arguments regarding the development of human history. Most importantly, he equated mathematics with contemplation, thereby establishing it as one of the first major steps towards true knowledge.

Nature imagery, such as fish swimming upstream, was used throughout this book. Such imagery, colourful and readily understandable, would have been expected by his audience and supposedly was easy for them to remember. An example of a different type of imagery which he sometimes employed was his notion that the Latins placed prudence in the heart.[11] Vico employed this and many other figurative expressions primarily as examples of the ways in which people with a limited vocabulary could express intangible concepts, but these phrases were also intended to demonstrate in a graphic manner various essential qualities, such as prudence and morality, which Vico deemed necessary for the proper functioning and survival – not only of early cultures – but of all human societies at each stage of their development.

Vico wrote that perception, judgement and reasoning were faculties peculiar to the human mind and that their regulation was by topics, criticism and method, areas of much less general interest to us now than to Vico, who was living in the final throes of scholastic tradition.[12] One might have expected him to have developed the discussion of these three faculties much further: he would have then been in line with his interest in the development of the human mind. Even in the first stages of mental development, always his prime concern, he recognised that these three faculties were all-important. Nevertheless he did differentiate emotional reactions and simple recognition from systematic forms of cognitive activity. He next proceeded to his distinction between knowledge, which was systematic and learned, and consciousness, which was intuitive.[13] This fundamental point was made first in *De antiquissima italorum sapientia* and was then maintained throughout the rest of Vico's writings. The distinction was significant for two reasons: it demonstrated that he considered there to be a difference between knowledge (*scientia*, which was systematic and learned) and consciousness (*conscientia*, which was intuitive); and it underlined his own emphasis on the intuitive and communal aspect of mental activity.[14]

Throughout *De antiquissima italorum sapientia* Vico discussed *phantasia* and *sensus communis* as the twin pillars upon which true intellect rested. His frequent references to mathematics, geometry and mechanics were used as concrete examples of the development of human mental faculties. In each of these fields the ability and desire to go beyond established limits was essential for further developments, particularly in mechanics, where the practical means to implement a workable project was as vital as the appropriate theoretical concept. This passing mention of the experimental method can be traced back to Bacon and *I Investiganti*. Vico identified ingenuity as an integral part of man's nature, the faculty proper to knowledge,[15] this being the case from childhood on and thus for societies throughout the development of a civilization. He discussed *ingenium* as gifts, talents, a type of mental power or activity. *Ingenium* was described as synonymous with nature, peculiar to man as a power which connects disparate things and as a faculty which develops with age as imagination diminishes with age.[16] He wrote that 'imagination was the eye of ingenuity but judgement the eye of intellect' ('*Phantasia ingenii oculus, ut iudicium est oculus intellectus*').[17] Creativity alone, then, was not true intellect, so much as it was the ability to recognize true worth, so-called constructive criticism, which was fundamental to intellect proper.[18]

Vico accepted that man did not have complete mastery over nature, even though he was a rational animal: 'God is the comprehension of all causes' (*'Omnium comprehensio caussarum est Deus'*) but, more originally he argued 'because man is neither nothing nor everything, he perceives neither nothing nor the infinite' (*'Homo quia neque nihil est, neque omnia, nec nihil percipit, nec infinitum'*).[19] This latter statement is a key declaration for Vico, for contrary to the generally accepted version of Enlightenment thought, he did not argue that man's potential for development was unlimited, but that it was restricted by lack of proper training, education and inclination. He asserted – and here there is a clear parallel with Hegel's and Marx's attitudes towards their own philosophies of history – that this realization was his first; but now, because of him, it belongs to all groups, not only to isolated individuals such as himself, and further that it was the responsibility of these recently enlightened societies to act in accordance with his historical insights.

Vico most certainly did not hold that man's mind was the apex of mental and rational development and that there was no need for the concept of a god. The desire in man to order his place in his relevant physical, social, mental and spiritual hierarchies was central to Vichian thought. However nowhere in his discussion of the limitations of the human mind did he state that social groups consciously recognised their abilities were restricted. This was his interpretation, not one he concluded came naturally from a study of the behaviour of societies.[20]

According to Vico metaphysics (*metaphysicus*, a theoretical basis) established the proper scope for each of the other branches of knowledge.[21] (He considered theology to be the most certain of all subjects.) Although a second-order discipline, as its role was to delineate other fields, metaphysics was in no way to be despised, for without it the other subjects could not function correctly. He likened the clarity of metaphysical light to that of sunlight,[22] and further used this analogy to state that physical objects exist in the dark.[23] Vico assigned to metaphysics roughly the same role in relation to all other academic subjects that modern interpreters have accorded his own contribution towards the provision of a means to approach the study of history – this being his critical (rather than speculative) approach to history. Vico was certainly not proposing a broad, all-inclusive interpretation of each of these various subjects – he wrote that to speak in universals was the practice of children and barbarians.[24] His main criticism of Aristotelian physics was that it was based on universals. He regarded the particular as always superior to the universal. Not only were ideas simpler to comprehend in an abbreviated form (examples and exceptions to set rules and standards, as already noted, were not unimportant to Vico as a teaching method), but for Vico universals do not give what is new, wonderful and unexpected. In a slightly different approach than in *De nostri studiorum ratione*, Vico stated in *De antiquissima italorum sapientia* that the best imitative artists are those who improve details, which once again one may interpret as a need to master universals before one can go beyond them, for only then can anything truly original be achieved. Vico's preoccupation with the idiosyncratic aspects of human life, manifested in *De antiquissima italorum sapientia* and throughout the entire *corpus* of his theoretical writings, lends considerable weight to interpretations of his thought which maintain that the goal of his study was the history of society, with all of its petty triumphs and manifest foibles.

4. *Il diritto universale* (1720–22)

Il diritto universale (*Universal Law*) is a substantial work by Vico, but to date it has been ignored by most Vico scholars. The first volume, *De universi iuris uno principio et fino uno* (*Universal Law, One Principle and One End*), published in 1720, is admittedly not generally of profound theoretical interest, grounded as it is in discussion and examples of Roman law. Nevertheless mixed in with his arguments concerning classical cases, he also mentioned key themes, often in the titles of his subsections: 'principles of all humanity' ('*Principium Omnis Humanitatis*'), 'true laws and certain laws' ('*Verum Legum Et Certum Legum*'),'customs and laws as expressions of law' ('*Mores Et Leges Iuris Naturae Interpretamenta*'), 'the natural order is the mind of civil society, the laws are the stories' ('*Ordo naturalis est mens reipublicae, leges sunt lingua*'), and the 'history of obscure times' ('*historia temporis obscuri*').[1] He wrote that governments must be preserved in the proper form in order to preserve the laws and that corruption of the laws always results when the natural order is forgotten. This book demonstrates how Vico would be attracted to a certain idea and term – for example, *ricorsi* (recourse) – and would use it many times before he clarified its definition in his own particular sense.

Nineteenth-century commentators, most notably Giuseppe Ferrari (1811–76), stressed the importance of Roman history on the development of Vico's thought.. More precisely it should be noted that Vico's emphasis was almost exclusively focused on Roman law. It was the history of Roman law, in the history of Rome, in the history of all nations, which was his aim. Likewise his autobiography was written as an individual demonstration of his general principles of world history. A chronological reading of his theoretical works allows one to examine the full formation of Vico's own ideas, to a much greater extent than with most writers.

The second book of *Il diritto universale* is entitled *De constantia iurisprudentis* (*On the Constancy of Jurisprudence*, 1722). It was followed by a book of notes (1722) and a final section of dissertations. Together they are now taken to comprise *Il diritto universale*. *De constantia iurisprudentis* is itself divided by Vico into two sections, the first of which dealt with the constancy of philosophy and the second, which begins with the chapter '*Nova scientia tentatur*' ('*New Science Attempted*'), with the constancy of philology. Etymology and philology were often discussed by Vico as being synonymous with historical reconstruction itself. He considered philology to be have two parts – the history of words and the history of human institutions ('«*Philologia*» *quid? Eius partes duae: historia verborum et historia rerum*'). This explains his allegiance to etymology (the study of words) and mythology (the study of happenings of the past). The seven and a half pages of '*Nova scientia tentatur*' are generally regarded – although seldom mentioned except in passing – as the rough draft for *La scienza nuova*, but the remainder of *De constantia philologiae* is itself an extremely useful and much neglected source of Vico's historical thought. It contains many ideas which changed little between 1721 and 1744, the publication dates of this second volume of *Il diritto universale* and that of the final version of *La scienza nuova*.

Vico's fascination with what he termed the 'magnificence of imaginings' ('*Imaginum granditas*') is one of the major themes running throughout this book.[2] He viewed imagination as the result of the poetic faculty of man, which disappears in the later phases of a society as the sciences gain in strength. Similarly he considered that where philosophy was robust, poetry diminished – for poetry originated through necessity and was the language of the first peoples.[3] He stressed over and over that original fables were produced by the least educated social groups. This was for him the 'poetic metamorphosis' ('*il Principio di tutte le metamorfosi, o sieno poetiche trasformazioni di corpi*').[4]

Giants are mentioned many times in his discussion of primitive man. Vico cited the development of a family structure and the fear of false religion, by which he meant the awe, the fear of the supernatural, inspired by any religion other than Christianity, as the two main reasons for their disappearance.[5] The obscure times of the giants were divided into smaller sub-sections by Vico, divisions easily recognisable from the various versions of *La scienza nuova*. In the earliest phase, that of divine authority, a theocratic society was the norm. Groups then progressed into a system in which authority was vested in families in the more general sense – extended families and kin groups. Then came the heroic, which Vico, rather confusingly, also called poetic, times. Changes occurred in these kingdoms with the diffusion of laws, in particular laws which protected the rights of the less powerful.[6]

According to Vico the first laws were born of the people and were not written down. Customs and ritual always preceded the development of laws and served as models for them; later the laws themselves were to function as the standard of proper behaviour. It was only the actual codification of the laws that distinguished them from customs.[7] Vico wrote that the first wise men were the poets who knew the laws, which were kept from the people.[8] He claimed, rather oddly, that even in Rome the jurisconsults served as the oracles for the people and poets of the city.[9]

Vico, then, discussed law in many different respects. Most importantly, he presented it as one of the best means of gaining historical insight into a past civilization. He especially prized ancient law, for it contained the purest form of a given language, thereby providing the necessary basis for a philological study of the even older form of the language from which it developed.[10] The vernacular used in classical Rome, for example, would not shed as much light on the origins of Latin as would the written form of the language. Hence the law also fulfilled an important role in his scheme as literature, as an expression of the creative spirit of a people. According to Vico 'poetic language is properly of religion and law' (*'Lingua poetica est religionis et legum'*), thus he included both spoken language and customs.[11]

Mathematics was once again stressed by Vico as an instrument of the application of practical knowledge, on this point in line with other thinkers of this period. He discussed geometry many times in this regard both in *De nostri temporis studiorum ratione* and in *De antiquissima italorum sapientia*. In the latter, for example, he cited the Egyptians' application of their mathematically-based astronomy to practical problems on earth.[12] Several parallels are drawn between mathematics and writing – he described the alphabet as the first geometry, and mathematics itself as the first writing.[13] It was of little concern to him whether the written word or mathematics developed first; he accepted them both as natural products of the developing human mind.

Rather unusually, at one point Vico veered away from his preoccupation with imagination and directly addressed the topic of human will. He argued that there were two origins of all knowledge, the intellect (*intellectus*) and the will (*voluntas*). He maintained that man was shaped by intellect and will in the same manner – that the consciousness of each is derived from either one, and that there cannot be one without the other.[14] Whatever Vico meant by this, he certainly did not hold that knowledge could be obtained in a passive manner. He occasionally dealt with knowledge as a capacity of a given individual – although this was always discussed in abstract terms – but more commonly as a stage acquired by a society; in this manner he defined knowledge as the necessity of reason, the arbiter of authority.[15]

For example, he wrote that at the beginning of knowledge, in the early stages of a society, the religious leader and civil leader were one and the same.[16] Hence it naturally follows that Vico also contended, predating Max Weber (1864–1920), that no society was founded on religious lines alone.[17] Vico praised the patriarchal society of the Jews in this book more for its effectiveness than for any particular spiritual virtues; in the same manner he stressed that the first mythology was a civil mythology.[18]

One of the most intriguing themes in this book – but one never directly addressed by Vico, only alluded to – is that of the development of the human mind. According to Vico when people are not vigorous in reason, the senses must be depended upon to a greater extent, just as nature has given animals very acute senses, and he continued (here in accord with Rousseau in *Émile*, 1762) that it followed that women were more sensible than men.[19] There is a very clear patterning of his objects of inquiry – he traced the development of human mental capacities through the senses, the growth of imagination and the fascination with magic and divination, as expressed by language and mythology. Set stages of mental development were at points spelled out (though they are more often suggested), but his often muddled examples demonstrate that he was not attempting to put forward a linear progression of human knowledge.

Nevertheless the discussion of language and poetry in *Il diritto universale* was firmly linked to Vico's growing realization of the need for some sort of historical method. Poetry and language were for him the natural starting points in his attempt to devise a method of reconstructing the histories of non-literate societies. He maintained that poetic language comprised primitive religion and laws, and that it was the late development of writing (relative to that of mythology) which kept most people unaware of their own history. Vico dealt for some pages here with the origin of language and the first human words – specifically with the order of appearance of the parts of speech – in a section not unlike the corresponding discussion in the final version of *La scienza nuova* (*1730, 1744*, Book II).

Poets were the first founders of literature, and it was the origins of heroic language or poetry which most intrigued Vico. He propounded two main causes for the ignorance surrounding the origin of poetry.[20] The first was the belief that the language of poetry was peculiar to the poets and hence different from that of the people. Perhaps no other single statement was so fundamental to Vico's interwoven fabric of language and history, here he was not at all concerned with the ruling elites of these primitive societies but with the structure and perceptions of the entire community. The second point concerned his conviction that the poets founded the false religions.[21] Vico does not dwell long on this issue, but certainly his intention was to reinforce the role of the poet simply as a medium, a human reservoir of the desires and fears of the people, not so much of his own generation but of those preceding him. Having established, at least to his own satisfaction, the inadequacy of previous approaches to the origins of language and poetry, Vico then proceeded to set out his list of *dignitate* ([Latin], *degnità*, [Italian] – axioms, a philosophical term, attributed in modern Italian dictionaries to Vico in particular) as the new basis for such an enquiry.[22] At this point Vico provided perhaps his most forthright statements concerning language and history: first, that language was inextricably linked with the expression of human creativity and second, that 'language gave a route to the mind, in regard to both civil and domestic customs, whether natural or moral, or domestic or civil' (*'nam linguae mentes solertes faciunt, cum ad quanquerem, sive naturalem, sive moralem, sive domesticam, sive civilem'*), and third that it was by means of language, and thus customs, that the nature of the past could be discerned.[23]

In response to the question 'what is history?' (*'Historia quid?'*), he answered conventionally: 'history is the witness of the times' (*'Historia autem est temporum testis'*).[24] The two-page chronology he presented at the beginning of *Nova scientia tentatur* was given as a background for his principles of universal history. The chronology itself varies little from the versions given in the various editions of *La scienza nuova*. He stated much more pronouncedly here than in any of his later and better known works that there is a double history – one of human institutions (*historia rerum*) and one of words (*historia verbum*).[25] Etymology and philology, he cited once again, were the appropriate means to investigate not only the history of words but the times in which they were in common usage.[26] He maintained that the origins and developments of past histories could be discovered through careful philological studies. Even more originally, he proposed that mythology could be used to penetrate the histories of what he often referred to as the 'fabulous times' (*temporis fabulosi*).

He considered secular histories to be of primary importance, since they were the means of entry into past successions of events. Many pages of this book are devoted to sacred history – the creation of the world, the Flood, the call of Abraham by God and the laws given to Moses on Mount Sinai. These were, according to Vico, the first four epochs of sacred history during the times in which secular history was for the most part obscure. Yet he remained preoccupied with the obscure and fabulous times, considering sacred history to be of interest primarily as a deviation from Gentile history. The overlapping of sacred and profane history held a peculiar fascination for him. As in some of his other works, notably *La scienza nuova prima*, Vico's enthusiasm for his topic takes him so far that he deemed it politic to retreat every few pages from the frontier between pagan and sacred history and to reiterate that antiquity demonstrated the perpetuity and truth of sacred history.[27]

There is no indication that when Vico spoke of history he was particularly keen to discover, verify or bring to life ancient civilisations, in the manner attempted by political and social historians of later centuries. There is no doubt that history, as the development of culture, was his primary concern. There is no evidence to suggest that he considered it possible to reconstruct, by means of his historical method, actual events of the past or the conditions in which these peoples lived. More importantly for an enquiry of this sort, Vico asserted that the perceptions and reactions of past peoples, not only to each other, but also to their social and physical environment, could be ascertained by means of philology (language) and mythology. It was without a doubt a history of ideas which Vico was attempting to construct.

For Vico the ultimate goal of such a study was a history of ideas. For all his discussions of Rome, he was not concerned with the production of a complete national history or even a unified history of ancient Rome or any particular social group. He was much more strongly motivated to bring to light common ideals shared by all people (which is discussed as *il dizionario di voci mentale* in *La scienza nuova prima*), and he speculated that this was a great deal easier to discern among primitive than modern peoples.

His interest in the development of human mental capacities demonstrated a strong desire to comprehend the workings and levels of the human mind in its more advanced state. He considered his proposed type of historical enquiry to be one way, if not the only way, to penetrate the gradual development of the human mind. Not only are ideas often all we have to examine of the past but at the most fundamental level it was ideas which motivated the actions and events of previous ages, regardless of whether these arose out of any conscious impulses. For this reason Vico's writings have an immediate relevance today. They not only point the way to what was in the eighteenth century a new method of approaching the past through language and mythology, but also they offer the very first elucidation of a new manner of viewing the past via the history of ideas, of *mentalités*. This second path, illuminated by Vico, is one which has only begun to be explored in recent times.

5. *La scienza nuova prima* (1725)

Vico's well-known failure to gain the Chair of Law at the University of Naples in 1723, which he had for so long desired, is generally accepted as signalling a major intellectual turning point for the philosopher. But this view does not take into account the unity of Vico's writings from 1699 onwards. Without a doubt Vico felt that his writing was most important, indeed it was all he had after this professional setback. By 1723 *Il diritto universale*, including '*Nova scientia tentatur*' (1720–22) had already been written, but the year after his defeat Vico turned to the writing of *La scienza nuova in forma negativa*, (*The New Science in Negative Form*; now lost). Max Fisch considers *La scienza nuova in forma negativa* to be parallel in its development to the first section of the autobiography. Fisch views *La scienza nuova prima* as aligned with the second section of the autobiography.[1]

In 1728 Giannartico di Porcía (who had organized the volume in which the autobiography appeared), and Antonio Conti (1677–1749), the patron of the same project, encouraged Vico to produce a new edition of *La scienza nuova*, but complaints from the Venetian printer regarding the repetitions and general unwieldiness of the revised work led Vico to publish the 1733 edition in Naples as he had the first. One cannot help but wish the Venetian publisher had had some influence on Vico.[2] In any case, the variations which make up this edition are now referred to as *Correzioni, miglioramenti, e aggiunte* (*Corrections, Improvements, and Supplements*) (1730)[3] The third edition, generally known, rather confusingly, as *La scienza nuova seconda* (*The Second New Science*), designates the 1744 edition plus passages from the edition of 1733, the corrections of 1730 mentioned above (but not the ones done in 1733), as well as the remaining sections of the 1733 edition.

There is no straightforward answer as to why Vico revised *La scienza nuova*. Encouragement from admirers of the book, his desire for a larger audience, plus the very probable genuine desire in his own mind to perfect these most important principles – all these were no doubt important factors. The reader familiar with the 1744 edition is forcibly struck by major differences when inspecting the first and second editions. The narrative style of *La scienza nuova prima* is a welcome surprise, entirely at odds with the list of *degnità* of *Il diritto universale*, of *Institutiones oratoriae* and the later, more popular versions of this work. Furthermore the 1725 edition is both more subtle and complete in its presentation of the major themes than the 1744 work. Perhaps there was an implicit assumption by Vico that most of the readers of his later edition would have read the earlier one – thus justifying his outline style in 1744 – or that readers new to his work would require his theories to be expressed more simply.

Why has the 1725 edition been so little used by Vico scholars? The 1744 edition is many times all that is cited in works on Vico – the two notable exceptions in the older scholarship of the English-speaking world being Adams and Berry.[4] Often the assumption is made that the final version contained the most complete and mature representation of his life's work simply because it was the last. This notion rests on the assumption of the internal coherence of a single writer's work, which derives in turn from the concept of progress. To be sure, questions regarding Vico's intellectual development from 1725 to the end of his life in 1744 are tantalizing. The academies and confraternities he frequented during this period, for example, would have given him at the very least a place to go and to talk, even if (as he repeatedly claimed) he did not have anyone with whom he could properly discuss his ideas.

The first book of *La scienza nuova prima* discussed the necessity of such a new science and the means whereby it could be discovered. But it is the second book on the principles of this science drawn from ideas, and the third, which discussed these principles as drawn from language, that are richest in Vico's highly stimulating notions regarding mankind, *il genere umano* (mankind). The fourth book, concerning the grounds of the proofs which establish this science, and the fifth and final one, on the philosophy of humanity and the universal history of the nations, are useful primarily as complements to the second and third book.

By way of contrast, the merits of the 1730 and 1744 editions are much more evenly spread. An explanation of the frontispiece, *'Idea dell'opera'* ('Idea of the Work'), begins the book, followed by Books I and II on the establishment of the principles and poetic wisdom. The small sub-section of seventeen lines on the Elements (*1725*, 208) was expanded to a rather large section of forty-three pages in the first book of the 1744 edition (*1744*, 119–329); it is this section which Fisch recommended as a starting point to readers new to Vico. Book III, *'Della discoverta del vero Omero'* ('On the Discovery of the True Homer'), is an addition to the later editions. It is very likely that the reason there is a more extended discussion of language in the first edition than in the others was that Vico had shifted his emphasis to the Homeric example; he viewed these epic poems as the classic example and supreme triumph of collective national wisdom, not of individual genius.

Book IV, *'Del corso che fanno le nazioni'* ('The Course the Nations Run'), expanded a section of the earlier work (*1725*, 400–1) and, in general, much more attention was given in the later editions to society and tradition. Although there is more discussion of Rome in *La scienza nuova prima*, in particular to Roman law, there is much less regarding the course of civilizations in general. The fifth and final book had the highly suggestive title of *'Del ricorso delle cose umano nel risurgere che fanno le nazioni'* ('The Recourse of Human Institutions which the Nations take when they Rise Again'). The short work, *'La pratica'* ('*Practic of the New Science*'), is now considered to be part of the so-called second *La scienza nuova*. *La pratica* offered a possible escape route from the return to barbarism which is implicit throughout *La scienza nuova* – that is, if the young are educated properly, in a society not already in decline, a people capable of maintaining public virtue will be created.[5] It has been surmised that this section was not published in his lifetime, because it implied that there was something external to his science that was needed to save nations from ultimate decline.

The complete title of *La scienza nuova prima* includes the concept, which was all-important to Vico, of the 'natural law of the peoples': '*Principi di una scienza nuova dintorno alla natura delle nazioni, per la quale si ritruovano principi di altro sistema del diritto naturale delle genti*' ('*Principles of a new science according to the nature of nations, in order to retrieve the principles of another system of natural law of the peoples*'). His definition of natural law was indeed broad, including as it did the origins of religions, language, customs, positive laws, societies, government, types of ownership, occupations, orders, authorities, judiciaries, penalties, wars, peace, surrender, slavery and alliances.[6] He viewed this natural law of the people as a 'jurisprudence of mankind',[7] believing one of its greatest strengths to be that it offered a method by which to analyse the barbaric stages of past civilisations. It offered a metaphysical explanation of the *certa mente comune* (certain common mind).[8] In addition Vico maintained that this natural law demonstrated the truth of the Christian religion. Hence natural law, according to Vico, was composed of virtually all elements of society, and yet it also proved (much less successfully) the validity of what may be viewed as the most controversial of its elements, that is, a particular religion and not religion in the abstract. Vico attacks Grotius, John Selden (1584–1654) and Samuel von Pufendorf (1632–94) for three reasons: they failed to note that natural law developed with the customs, they did not give us a way to find out the history of barbarous times and they do not discuss individual nations and cases, only common traits of all nations. Vico began this work by declaring that natural law developed with the customs, but within a few pages he used the terms synonymously. Vico had no patience for the universal approaches of the three thinkers mentioned above who dealt with politics, war and foreign relations to the exclusion of society, culture and history. He demanded knowledge of what was unique concerning each society, and had no time for universal theories, which provided no means of examining the particular.

At the very beginning of *La scienza nuova prima*, Vico condemned curiosity about the future as irreligious, but elsewhere in his writings he used the same term, *curiosità*, only in a positive, constructive sense. Vico wrote that it was curiosity that led early man to explore his environment and, understandably, to judge the unknown by the known.[9] This issue of curiosity is central to Vico's ideas about language, the development of the human mind and history itself. He was persuaded that only when the desire to push back established boundaries of knowledge and learning was present could additional knowledge or insight be gained. Vico would not have argued that this desire had to be fully articulated or even conscious, simply that the demand for additional information was in itself the vehicle by which the human mind progressed.

Both editions make repeated references to the dictionary of mental words, *un dizionario di voci mentale comune a tutte le nazioni* (a dictionary of mental words common to all the nations), a marvelous phrase which Vico concocted for this rich and diverse group of innate ideas. Unfortunately he did not list them in the last edition in as specific a form as in the 1725 edition, which proves rather confusing for one who has read the more popular version.[10] The first edition expressly listed the twelve somewhat disjointed concepts – not single words as the following summary indicates – which all societies shared: imagined deities, children begotten under divine auspices, beings of heroic origins, divination, sacrifice, complete power over families, the strength to kill wild animals and to cultivate and defend the land, magnanimity towards the poor, the fame brought by ability to crush enemies, sovereign dominion over the fields, the power of arms and the power of law.[11] These twelve concepts or categories were viewed by Vico as the minimum requirements, the fundamental practices of all societies; he went so far as to consider them necessary ingredients of all early cultures. Only the point regarding magnanimity towards vagabonds did not fit naturally into his list – it seems to have been a rather gratuitous addition, perhaps made to stress the civilising aspects of a developing society.

This concept of a *dizionario mentale* was fundamental to Vico's views regarding history. His *storia ideale eterna* and natural law could only work if all people shared some common goals, regardless of the time, place or circumstances of birth. This concept led naturally to his assertion that two civilisations could develop in a similar pattern and share common beliefs without any direct contact occurring between the two. The notion of originality here seems to have been turned on its head by Vico. It is commonly assumed that whoever was first to devise an idea or approach is to be credited with this breakthrough and that all subsequent work in this area will go on from this point, even if only as revision. But Vico seemed to have considered it neither meritorious nor even possible to conceive a new idea, because there was for him – at least in the abstract – no such thing. For Vico it was the application of these ideas which was the most decisive issue.

One finds in *La scienza nuova prima* in five-point form perhaps the clearest statement in any of Vico's writings regarding his methods of history. The first is the use of evidence synchronous (*sincrono*) with the times in which the Gentile nations were born.[12] Unfortunately for us, Vico did not develop this point, but one can safely assume tradition to have been a major aspect. The word *sincrono* would allow for a broad range of Vichian subjects. For example, the development of language, legends regarding the deities and speculations regarding the creation and workings of the natural world formed a natural part of this approach.

In his second point, Vico cited the use of documents belonging to the first peoples, and stated time and time again that men naturally preserve records of their past history.[13] His primary example, and one still studied today, is that of independent verifications of the Flood in the records of past civilisations which were not in direct contact with one another. The use of documents was not an issue which Vico stressed, surely because its importance was so obvious. Yet it was in regard to the examination of such documents that Vico gave his well-known warning that one should not suppose people in the past to have been better informed than ourselves about times that lay closer to them and he mentioned forgeries in this connection.[14]

The third point concerned what Vico chose to call physical demonstrations (*fisiche dimostrazioni*).[15] Giants and monsters were for Vico physical proof of the bestial stage of Gentile man left over from a Hobbesian state of nature.[16] Scattered references to these *bestioni* are to be found particularly in *De constantia philologiae* and in the last edition of *La scienza nuova*. It is not clear why Vico felt giants be a necessary part of his scheme, for he does not discuss the physical development of human beings. Understandably, this has been interpreted as a proto-evolutionary view on Vico's part. Yet nothing would have horrified him more, and it could well be argued that it is indeed superfluous to Vico's main argument regarding the development of the human mind.

The fourth means Vico listed concerned the rational proofs contained in fables. He did not, however, consider this method by any means simple or guaranteed to work. He gave the examples of contemporary idiots and feral children, citing the difficulty in establishing any sort of intellectual contact with them, despite their having had some access to modern civilization.[17] At one point Vico wrote that fables were fictional expressions of tangible objects; but elsewhere he expanded this definition to include substantives, since emotions and ideas were often personified in mythical form.[18] First speech, he wrote, was an outburst of such passion – interjections.[19] He spent a great deal of time discussing the origin of the poetic character which constituted the vocabulary of Gentile nations and the meanings of true poetic allegories. For Vico language went hand-in-hand with fables, myths and poetry, and together they were a testimony of the ancient peoples.[20] Vico was most explicit in his discussion of myths and fables in *1725*. Myths must be a credible impossibility, they must inspire awe and fear, and they must contain a worthy message.[21] In the 1744 edition this three-fold aim was slightly changed to encompass the invention of sublime tales, the inspiration of awe and fear (exactly as in *1725*), and the teaching of the vulgar to act piously.[22] The final point, important in regard to the implications of '*La pratica*', was mentioned in the 1725 edition but had not been ranked as one of the goals.

In the fifth and final point Vico contended metaphysical proofs (theoretical arguments) to be necessary when the other four means proved inapplicable or inadequate.[23] Vico asserted that he had discovered the origins of idolatry and divination in this manner.[24] To some extent, Vico's entire schema – including its more tangible aspects such as language – could be grouped under this metaphysical heading. He did not speculate merely about the decline following the development of a nation, but also about areas now considered analytical in the philosophy of history; and it is not unreasonable to surmise that it is his achievements in the latter area which will be of lasting interest. Vico did indeed draw up a framework of past civilisations – the most tangible example being *La tavola cronologica* (the chronological table), which appeared in all three editions.

The bulk of each edition is devoted to a quite different issue – how *entrare* and *descendere* into the minds of these *grossi bestioni*, and how then to interpret this information concerning the past. Imagination, human creativity, is the theme which runs throughout this book. Vico avowed that he had formulated both a history and a philosophy of the law of mankind. His principles of mythology and etymology were to be used to retrieve the vocabularies, and hence the mentalities, of early societies:

Finalmente il niuno o poco uso del raziocinio port robustezza de'sensi. La robustezza de'sensi porta vivezza di fantasia. La vivida fantasia è andl'ottima dipintrice delle immagini, che imprimono gli oggetti ne'sensi.	Finally an absence or scarcity of reasoning brings with it a strength of the senses which, in turn, leads to a vividness of imagination, and a vivid imagination is the best painter of the images which objects impress on the senses.[25]

Vico has been credited with so many original insights that it is important to distinguish those which are truly his from those which have been falsely attributed to him, many of which are not only unrelated but are diametrically opposed to his principles. Falling into this last category is the notion of freedom, either in the abstract or as a specific goal for individuals or societies.[26] This would seem to be not unrelated to the issue that Vico was not particularly interested in the role of the individual. Although very much concerned with the creative faculties of man, his interest lay in the results of collective inclinations, not in individual achievements. Thus, anti-social behaviour, which one might have expected to intrigue Vico as a break from his ideal, eternal pattern, is ignored in his writings as insignificant; or perhaps he simply deemed it a natural part of any, and thus all, societies. His mention of Julius Caesar or of the need for an Augustus notwithstanding, Vico's views most closely approximate those of Hegel on 'World Historical Individuals' ('*die weltgeschichtlichen Individuen*') on this point.[27]

Although a strictly deterministic application of Vico's rise and fall of civilizations robs his work of much of its flavour and diversity, it must be acknowledged that he never denied the fundamental role of this pattern for his outlook on human history. Yet as Vico himself often repeated, it was not those eternal truths that transcended any one particular society which most tantalized and inspired him, but the process of investigating these past social groups by means of language, mythology and imagination. These were the factors which constituted the major components of Vico's highly original – and still much misunderstood – approach to the past.

6. *Vita di Giambattista Vico scritta da sé medesimo* (1725, 1728)

The original impetus for Vico's autobiography came from Gottfried Wilhelm von Leibniz (1646–1716). Count Giannartico di Porcía, a Venetian nobleman, learned of the German philosopher's proposal for an anthology of contemporary intellectual autobiographies. Thus inspired, Porcía asked Vico, along with eight other Neapolitan thinkers – including Paolo Mattia Doria (1661–1746), Vico's friend, and Pietro Giannone, (1676–1748) his rival – to write their autobiographies, which were to be published together in a single volume.[1] In the event, Porcía persuaded only Vico to contribute, no doubt because he never completely gave up hope of achieving the sort of wider recognition which was to elude him in his lifetime.[2] Porcía proposed to publish Vico's work by itself as a model for future autobiographies, but Vico was quite rightly afraid that he would be mocked for writing such an intensely personal piece and only very reluctantly gave his permission for its publication two years later. Indeed Giannone, the famous anti-clerical writer of Naples, called the autobiography, when he read it in Vienna in 1729, 'la cosa píu sciapita e trasonica insieme che si potesse mai leggere', ('the most insipid and dreamy thing that one could ever read').[3]

Written in Italian, the original section of the autobiography is over fifty pages in length. Four other sections are now considered to be part of the autobiography as well.[4] In 1730 Vico was elected to the Academy of the Assorditi of Urbino, through the efforts of the well-known Neapolitan historian, Ludovico Muratori (1672–1750), on his behalf. The Academy asked for a biographical piece, and, not wanting to renew the conflicts surrounding the publication of his autobiography, Vico simply revised it as *Aggiunta fatta dal Vico alla sua autobiografia* (*Supplement made by Vico to his Autobiography*, 1731). The first part (1725) of the original section of the autobiography was completed just after *La scienza nuova in forma negativa*, and the second section (1725, 1728) after he completed the first edition of *La scienza nuova*. Further, the second *La scienza nuova* displays a number of parallels with the 1731 addition to the autobiography.[5] Just as Vico added the numerous classical examples (especially regarding Homer) to the later edition of his book, he expanded the second section of the autobiography with numerous quotations from letters sent to him by people who had received (usually unsolicited) copies of his book.

The Marquis of Villarosa found a copy of *Aggiunta fatta dal Vico alla sua autobiografia* among the papers of Vico's son, Gennaro, in 1806.[6] Villarosa himself wrote *Gli ultimi anni del Vico* (*Vico's Final Years*, 1818) based loosely on oral tradition, it is now considered to be the third part of the autobiography. Villarosa's study of Vico's last years is tragically concise: his unhappy family life; small salary; frustrated academic career; a wife with whom he had very little in common; one son a criminal; senility which struck him at least fourteen months before his death; and the physical fight over his corpse between the members of the Confraternity of Santa Sofia and the professors of the university. Villarosa considered this to be literally the final insult to Vico. Happier entries might have included the general high regard in which he was held in the various academies he frequented; his relationship with his daughter, Luisa, who became a poetess of note in the city; and the professional success of his son, Gennaro, who succeeded him in his Chair of Rhetoric. Although Gennaro was perhaps too helpful to his increasingly senile father during his final years, and might be responsible for encouraging the generally prosaic changes made between the 1730 and 1744 editions of *La scienza nuova*.

The fourth and final part of the autobiography, *Due appendici* (*Two Appendices*), contains the *Cataloghi delle opere del Vico compilati dall'autore* (*Catalogue of Vico's work compiled by the author*, 1728, 1735) and *Le recensioni di Giovanni Leclerc tradotte e annotate dal Vico. Notizie sparse e documenti per la vita del Vico* (*A few notices and documents relating to the life of Vico*) is the final section of the autobiography. Villarosa, then, was responsible for three of the five sections, having written the third and collected the materials which comprised the final two.

It is generally accepted that Vico used Descartes's *Discours de la méthode* as the model for his autobiography. There is some sense to this view, since the *Discours de la méthode* was largely autobiographical in approach; nevertheless, it is somewhat ironic that by 1725 when Vico wrote his autobiography he was undoubtedly anti-Cartesian. The use of the third person singular by Vico is seen by Max Fisch as a response to Descartes's 'ubiquitous "I"'.[7] But other considerations should perhaps be taken into account: the now familiar first person autobiographical style, for example, had not been established at this time, and it must also be noted that the voice of a supposedly impersonal narrator allowed Vico to flatter his own accomplishments and justify his failings. Seemingly endless quotations are given of people praising Vico, particularly in the second section, so many that sadly one wonders if every compliment he ever received is recorded here. In Vico's defence it must be stated that by 1725 he may have realized, quite rightly as it transpired, that if he did not take the opportunity to record his own biographical details and to present his work to a wider audience, it would never be done. Indeed later biographical sketches of Vico (including this one) are almost entirely dependent on the autobiography since there are few other sources to consult. There is no self-criticism of his life or work in the autobiography, only justifications for obvious failures. Clearly there is a need to apply gracious Vichian modes of criticism to Vico himself, if we are to avoid too severe a judgement.

From the autobiography one gains the sense that Vico was less than proud of his family background. His father, Antonio di Vico, had a small bookstore in Naples. His mother, Candida Masullo, was the daughter of a carriage maker, and is generally believed to have been illiterate, as was his wife. Porcía had asked that the contributors to his proposed anthology discuss their '*tempo della nascita*' (time of birth), the '*nome de' loro Padri e della loro Patria*' (name of their fathers and of their countries), and '*tutte quelle aventure della loro vita, che render la ponno più ammirable e più curiosa*' (all of the adventures of their lives, which are most admirable or most curious).[8] However Vico omitted the names of his parents and claimed to have been born in 1670 rather than 1668. It is not altogether clear why Vico felt compelled to make himself two years younger, as he was the sixth of eight children born to his father's second wife and so was presumably free of any taint of illegitimacy. Nicolini was perhaps correct in speculating that it had to do with Vico's embarrassment over the long interruption to his schooling caused by a fall on his head as a child.[9] Vico was a precocious child and was treated warily by his teachers at both school and university. His true ability was never recognized at any stage of his life.

It is from the autobiography that the best known facts regarding Vico's life are to be found: this fall on his head as a child, his failure to gain the prestigious chair of the head morning lecturer of law in 1723, and the sale of his ring to pay for the printing of a shortened version of his work, *La scienza nuova in forma positiva* (*The New Science in Positive Form*, 1725). Vico stressed the critical importance of the nine years he was away from Naples in preserving him from the corruption of Cartesian thought which he claimed gained prominence in Naples during this period.[10] The years in question (1686–95) were spent as a tutor to the son of Don Domenico Rocca at the family estate in Vatolla, approximately one hundred kilometres south of Naples: many years later the son participated in the 1701 conspiracy of the nobles, of which Vico was to write an account. Vico's own poetry, written at the age of twenty-four, perhaps for the daughter of the house, *Affetti di un disperato* (*Emotions of a Desperate Man*, 1692) offers no clues to his theoretical works on language and poetry, although it is of interest when studying his intellectual development because of the Lucretian overtones.[11] These were the only years Vico spent away from Naples, and thus they are of some particular interest when studying his intellectual development. Still there is little doubt that Vico overestimated the importance of this period in terms of his escaping the snare of Cartesianism, particularly because the Cartesian spirit was already very much alive in Naples among the generation preceding Vico – Leonardo di Capua (1617–95) is one exponent of this. There is also evidence to suggest that the entire Rocca family spent several months of every year in the city of Naples, hence Vico in his capacity as the son's tutor would have been back in the city at regular intervals and thus would have been exposed to the prevailing intellectual currents himself. Nevertheless his claim must be taken seriously, for Vico used it – consciously or not – as a means of separating himself from contemporary intellectual trends, from Naples and all that it represented.

Although later writers are most concerned with Vico's philosophical differences with Cartesian thought, this went along with his horror regarding the diminished status of the sixteenth-century Italian neo-Platonic writers (many of whom he had avidly read while at Vatolla), due in large degree to the vast and still-growing influence of Cartesian thought. Ironically Vico's insights into the interpretation of poetry are now considered to be the apex of the early eighteenth-century Italian school of literary criticism which he himself loathed.[12]

There is no doubt that Vico was to a great extent removed from the mainstream of intellectual life in Naples. Not only did he not achieve the kind of professional success which he so keenly desired, but he had little in common with the two most dynamic groups in Naples which overlapped the opposite ends of his lifetime, the scientifically-minded *I Investiganti* and the economic and political writers of Genovesi's circle. Vico's strange inability to read either French or English, or indeed any modern lanaguage except Italian, which was not at all typical of other Neapolitan intellectuals of his generation, cut him off from much of the more exciting work produced at the turn of the eighteenth century. These circumstances further tied him to the seventeenth century: when Vico boasted that he never read a new book after 1709 – at the age of forty-one, thirty-five years before his death – one is not inclined (along with Nicolini) to believe him. Certainly Vico was familiar with the work of Jakob Brucker (1696–1770) on philosophical eclecticism.[13] Nevertheless, the date when Vico supposedly stopped reading new books, 1709, takes on a greater significance when it is realized that that was the year that Vico published *De nostri temporis studiorum ratione*. Although Vico never mastered another living language, in an academic and abstract manner he was greatly interested in the development of language and philology, and language played a fundamental role in his theoretical works regarding the growth of human creativity and imagination.

Much of the recent work on Vico, especially that done on his theory of knowledge, considers him to have secretly harboured heretical views, and that his purpose in using allegory was to attack Biblical authority. Homer, according to this view, was used as a synonym for Moses in '*Della discoverta del vero Omero*' in much the same manner that the central character in Voltaire's *Mohammet* (*Mohammed*, 1741) was popularly inter- preted as Christ. This position is supported by existence of the Inquisition in Naples during Vico's lifetime and Vico had personal acquaintances called before it.

Yet many Neapolitan historians, of which Croce is the best known, have stressed the Neapolitan loathing for the Inquisition and its subsequent inef- ficacy. Giannone's sharp attack on the power of the papacy in *Dell'istoria civile del Regno di Napoli* (*On the Civil History of the Kingdom of Naples*), which was quickly placed on the Index after its publication in 1723, was not out of line with the anti-clerical mood of many influential Neapolitans of their generation.[14] Throughout this period the Kingdom of Naples was engaged in official disputes with Rome over papal claims to Naples dating back to 1053 and 1059, and in the third quarter of the eighteenth century Bernardo Tanucci (1698–1783) and Ferdinando Galiani (1728–87) formed a successful alliance to drive the Jesuits out of the Kingdom of Naples altogether.[15] Nonetheless there is no evidence in Vico's autobiography or elsewhere to suggest that he personally found the Catholic Church to be a stumbling block; quite to the contrary, he recounted with great pleasure a number of conversations he had had both as a student and as an adult with intellectual priests, several of whom he counted as friends. In this century Catholic writers prefer to speak of Vico's 'sweet, Christian manner'; the late eighteenth- and nineteenth-century anti-Vico controversy among Catholic scholars, exemplified by Finetti, has not been revived.[16] Yet precisely because many of his ideas were so startlingly modern, there remains a tendency for subsequent generations of readers, who assume that Vico's work was all of a piece, to regard his opinions as being in full accord with their own (often, in reality, quite contrary) views. When, in reality, it might well be argued that Vico was unaware, and indeed would have been horrified, by many of the later conclusions drawn from his own ideas.

Vico's formal education was not unusual for his time, but here again he stressed his own individuality (somewhat ironically, as he had no theoretical interest in individuals, only in social groups) by blessing his great good fortune in not having become the protégé of any one of the town's more forceful intellectuals, thereby leaving him untroubled by the fighting among the various academic factions.[17]

Talché per tutte queste cose il Vico benedisse non aver lui avuto maestro nelle cui parole avesse egli giurato, e ringraziò quelle selve fralle quali dal suo buon genio guidato aveva fatto il maggior corso dei suoi studi senza niun affetto di setta, e non nella città nella quale, come moda di vesti, si cangiava ogni due o tre anni gusto di lettere.	So for all these reasons Vico blessed his good fortune in having no teacher whose words he had sworn by, and he felt most grateful for those woods in which guided by his good genius, he had followed the main course of his studies untroubled by sectarian prejudice; for in the city taste in letters changed every two or three years like styles in dress.[18]

It was his prodigious private reading which first distinguished Vico from his peers. He claimed to have read classic works without the use of any commentaries three times: once for an understanding of the composition as a whole, next to note the transitions, and finally to trace the development of the ideas.[19]

Although he discussed his *quattro autori* in some detail in the autobiography, these four writers had both a symbolic and a tangible effect on his own writings – they are sometimes viewed as an 'allegory regarding his own intellectual development, a sort of private myth', as Peter Burke expresses it well.[20] In any case, Vico was influenced much more by neo-Platonic writers than by Plato. He relied on Tacitus and Grotius in his work on Caraffa. Tacitus's *Germania* (98 A.D.) was crucial in shaping Vico's own concept of primitive society, and Vico shared Bacon's fascination with myth. There is certainly a parallel between Bacon's stress on reproducing nature in a laboratory setting and Vico's idea that one can only know completely what one has made; although Vico used this argument to state that one can never fully understand nature.

Vico discussed each of his own main works and their various editions in chronological order in the autobiography, but little new light is shed on these pieces other than the circumstances under which they were written. His presentation of his own writings led up to *La scienza nuova* in a straightforward chronological manner, but, significantly, he did not treat *La scienza nuova* as his only work of importance. A great many questions are left unanswered by the autobiography. There is, for example, unfortunately no explicit statement regarding his increasing preoccupation with history and cultural development. Yet certain ideas and terms recur throughout this work – *lingua*, *memoria* and *fantasia*, and *ingenio* (ingenious), a term he often used to describe himself. In the autobiography, as in *Il diritto universale*, he used his favourite terms over and over, whether in an original form or not. His obsession with particular ideas is evident. The continuity of the development of his ideas is clear from the autobiography, and this development is equally clear from his writings.

The real importance of the autobiography, then, has little to do with the sometimes muddled facts which Vico recounted regarding his life. Its true significance lies in the fact that it was one of the first examples of an original writer's examination of the sources and origins of his own ideas. There was a direct parallel, for example, between his autobiography and his view of myths, not as false statements about reality or fanciful versions of past events but as evidence of early outlooks and beliefs. Hence Vico's autobiography was written not only as a model for future intellectual autobiographies but also as a practical lesson in the application of his own view of history.

3 *La scienza nuova*, 1725, 1730, 1744

1. The Three Editions

Vico wrote three versions of his last work, *La scienza nuova*, all of which were published in Naples in 1725, in 1730 and in 1744, the year of his death.[1] With very few exceptions, it has been the final version which has been considered the definitive edition, to the exclusion not only of the two previous editions but also of all of Vico's other writings. Only in this century has this overemphasis been somewhat rectified in the Italian-speaking world by research on Vico's early works, especially in their specific Neapolitan or Italian cultural contexts,[2] and in addition, more recently, by the 1981 publication by Lessico Intellettuale Europeo of a concordance of the 1725 edition[3] (with concordances of the rest of his works to follow shortly) and the release in 1987 by the Centro di Studi Vichiani of the orations as the first volume of the long-awaited scholarly edition of Vico.[4]

The difficulty has not only been the lack of modern, scholarly editions of Vico's works, nor that the 1744 edition is generally the work first to be translated;[5] the stumbling-block has been the assumption by generations of readers that the edition they were reading was the best because it was the final one and thus Vico must have considered it the best. It is true that Vico wrote in 1731,

A' quali per far loro verdere che gli conosceva quali essi eraano, fece intendere che di tutte le deboli opere del suo affannato ingtegno arebbe voluto che sola fusse restata al mondo la *Scienza Nuova*, . . .	To show them that he knew them for what they were, Vico gave them to understand that of all the poor works of his exhausted genius he wished only the *New Science* to remain to the world; . . . [6]

The edition he was referring to was the 1730, and the only exception he made in this regard was for the essential section in the 1725 edition on *i voci mentale*. At this point he had not yet written the work, the 1744 edition of *La scienza nuova*, which most interpreters claim is the quintessential Vico.

Contrary to what some modern interpreters claim, Vico did not write that he considered the 1744 edition the definitive one (as it had not been written at this point), but that the 1730 edition plus this one important section from the 1725 edition were to be his lasting memorial.

There is another reason for the dominance of the final edition, which is that the interpretations of Vico championed by Michelet, Croce, Collingwood and Berlin have been so strong, and the interest in Vico generally so slight, that many established tenets of Vico studies have never been questioned. Such an attitude would never be tolerated in studies of better-known thinkers of the time, such as Descartes or Voltaire.

This complaisance is nowhere more obvious than with regard to the issue of the various editions of *La scienza nuova*. In the previous chapter the 1725 edition was discussed in the context of Vico's most significant early works. The issue of the *dizionario mentale* never again received such full treatment as in the first edition.[7] Yet Vico's concepts, not only of imagination but also of history, received only passing mention in *La scienza nova prima*, hence the 1725 work could never be considered the definitive edition.

One of the many problems with the final edition concerns the long passages in which Vico supposedly applied or demonstrated his theories by references to classical history or classical and popular mythology, which he supposedly hoped would increase his readership. While it is no longer possible to judge the exact reaction of Vico's contemporaries to his literary style, it is well known that such references were *de rigueur* in eighteenth-century theoretical works. One must imagine that it was not these classical allusions, which the modern reader finds so tiresome, but Vico's own unconventional ideas which were responsible for the poor sales of his works in the eighteenth century.

Long, tortuous passages on Roman law and heroic figures, not to mention his many misquotations and spurious allusions (faithfully rectified by Ferrari, Pomodoro, Nicolini and later editors)[8] form the bulk of the final version. These additions were written in 1730, 1731 and 1733 and were entitled by Vico *Correzioni, miglioramenti, e aggiunte*.[9] Sadly the 161 folio pages of closely written corrections written in Vico's own hand add very little to our understanding of his thought, for they are almost entirely concerned with the Roman and mythological examples to be found in the 1744 edition.

This is not to say that *La scienza nuova* was unimportant to Vico's thought, but that we have not been reading the correct edition. It is neither the 1725 nor the 1744 edition to which we should turn our attention, but to the 1730 edition – particularly in regard to imagination. This all-but-forgotten edition, which even Nicolini abridged, has never been republished in its entirety. The sections of Vico's work that are of least interest to modern readers – classical mythology in Books IV and V of the 1725,[10] and the classical and mythological references of the 1744 – form only a small portion of the 1730 edition. Certainly the discriminating reader is capable of skimming or passing over less interesting passages. But the issue is more fundamental than this. In the 1730 edition Vico devoted most of his energies to the topics in which he was to provide the most interesting and original insights: language, imagination and history.[11] One of the only previous supporters of this view was Croce:

> His system is recapitulated and brought to perfection in the *Principî di una scienza nuova intorno alla commune natura delle nazioni* (1725) and especially in the second *Cinque libri de' principî di una scienza nuova* (1730, new ed. 1744).[12]

Virtually all of the 1730 edition is repeated in the 1744 (some of the few exceptions are cited at various points throughout this book). Nevertheless, if one must choose a definitive edition of *La scienza nuova*, based on its own intrinsic merits rather than on the grounds of availability (which, granted, is no small consideration for those not able to consult one of the very few extant copies – the manuscript has not survived), then the 1744 edition clearly is inferior to that of 1730. This view is by the two modern Italian scholars who have spent the most time studying the Vico manuscripts and first editions – Manuela Sanna, who compiled the *Catalogo vichiano napoletano*[13] in 1986 and Paolo Cristofolini, who edited the excellent Sansoni editions of Vico's *Opere filosofiche* in 1971 and *Opere giuridiche* in 1974[14] and who is editing a complete, scholarly edition of the 1730 edition, expected to be published in the near future. The 1730 edition reads like a first-class précis of the often incomprehensible finished product of 1744. It is to be hoped that after more than 250 years of near total neglect, the 1730 edition will at last receive the attention it so clearly deserves. This streamlined version would be more likely to attract modern readers than the other two editions. In the future Vico studies should concentrate on the 1725 and 1730 editions of *La scienza nuova*, in conjunction with his earlier writings, in order to gain the most complete representations of his ideas.

2. Vico's Vocabulary

As we have seen, most previous studies have restricted themselves to an analysis of the third and final version of Vico's best known work, *La scienza nuova* (1744). Little effort has been made to trace the development of his ideas in the important first (1725) and almost unknown second (1730) editions of *La scienza nuova*. It is in the second edition, never reprinted in its entirety, that he devoted most of his energies to the topics in which he was to provide the most interesting and original insights – language, imagination and historical knowledge. (Most of these ideas were repeated in the final edition but the emphasis there is on Roman history, classical mythology and historical cycles.) Yet a study of these three crucial topics in Vico, each of which will be dealt with in greater depth in later chapters, must be firmly grounded in a study of the many diverse terms and phrases which comprised Vico's own very idiosyncratic vocabulary.

According to Vico, the best method to search out early human institutions was to study the mental vocabularies of these times.[1] It was the poetic characters, the gods, whom he believed comprised the vocabulary of the Gentile nations.[2] Although he claimed to be well aware of the difficulties in such an historical reconstruction,[3] he argued that it was impossible for these peoples to have created false ideas, because there was no tradition which did not have some basis in truth – the ideas could not have been completely manufactured.[4] Just as Vico asserted it was necessary to break the poetic codes of early, preliterate societies in order to discover the mentality of those times, in the same way we must break the code of Vico's own vocabulary. There are seven main concepts which must be addressed: The notion of the uniformity of ideas in all societies at the same stage of development, *sapienza volgare* (vulgar, common, sometimes called poetic wisdom), religion and society of these early groups, the relationship between free will and the development of society in Vico's thought and, finally, Vico's division of human history into *setti di tempi* (periods of time) which can be examined only by means of his new critical art.

3. Uniformity of Ideas

The third book of the 1744 edition, entitled '*Della discoverta del vero Omero*'[1] was based on an earlier two-and-a-half-page draft written in 1728–29, entitled by Nicolini, '*Della discoverta del vero Dante*' ('On the Discovery of the True Dante').[2] The shorter piece listed the three primary reasons for reading the *La commedia divina* (*The Divine Comedy*, 1307?–1321), which show a striking similarity to Vico's later arguments for studying early poetry and mythology, and hence history. The first reason was that *La commedia divina* itself was a history of the barbarous times in Italy. This is a clear statement of Vico's conviction that it was from oral history, and later codified versions of the same, that one could gain historical knowledge of the past. Sixteen years later he wrote in his best-known work:

Ma sopra tutto, per tal discoverta, gli si aggiugne una sfolgorantissima lode: d'esser Omero stato il primo storico, il quale ci sia giunto di tutta la gentilitá; onde dovranno quindi appresso, i di lui poemi salire nell'alto credito d'essere due grandi tesori de'costumi dell' antichissima Grecia.

But above all, in virtue of our discovery we may ascribe to Homer an additional and most dazzling glory: that of having been the first historian of the entire Gentile world who has come down to us. Wherefore his poems should henceforth be highly prized as being two great treasure stores of the customs of early Greece.[3]

Thus for Vico the primary importance of poetry was the evidence it contained of past cultures, in this manner he believed Roman law to be a 'serious poem'.[4] *'Della discoverta del vero Dante'* is further evidence that, long before 1744, Vico considered early poetry to be crucial not only as literature but also as philology and history. He postulated that Dante's (1265–1321) masterpiece was created by shared Italian wisdom. Vico himself attempted, unsuccessfully, to emulate this *'puro e largo fonte di bellissimi favellari toscani'*[5] ('pure and large fount of the beautiful Tuscan language') by modelling his own writing style on the Cruscan model.[6] But more importantly, he prized the pure form of any language for its invaluable role in philological studies. Although attention is now quite rightly placed on Vico's interpretations of mythology, rather than on the academic methods by which such a reconstitution could be accomplished and which he never discussed in a specific manner, Vico himself repeatedly stressed that philology and etymology were virtually synonymous with historical studies.

Vico declared *La commedia divina* to be an example of sublime, lofty and majestic poetry. Yet it seems clear that one of the main reasons Vico shifted his example from Dante to Homer was that he wanted to discuss mythology rather than the work of one man, however exceptional. Book III of the 1744 edition made clear that Vico regarded the works of Homer to be the result of the collective wisdom of the Greek people.

In cotal guisa si dimostra l'Omero autor dell' *Iliade* avere di molt'etá preceduto l'Omero autore dell'

In this fashion we show that the Homer who was the author of the *Iliad* preceded by many centuries

Odissea.   the Homer who was the author of
                                        the *Odyssey*.[7]

Not only the discussion of Homer, as with that of Dante in the earlier
'*Della discoverta del vero Dante*', but also in *La scienza nuova prima*
(three years previously) there was a strong argument that many of his
conclusions were leading in the same direction as his discussion of Moses
in 1725.[8] If Vico himself was aware of this possible interpretation – that
is, that Moses was not an historical person, but the personification of the
wisdom of the ancient Jewish people – it would no doubt account for
his twice changing his principal example, from Moses to Dante and then
from Dante to Homer. This argument need not depend on any supposedly
heretical views which Vico wanted to hide, since the issue could have
been simply that he did not want his ideas to be interpreted in such
a way. For, as already mentioned, Vico did not even exhibit the usual
anti-clerical attitudes rife among Neapolitans of his time, attitudes which
culminated after his death in the expulsion of the Jesuits from the Kingdom
of Naples.[9]

Vico's shift from Moses to Dante to Homer is most intriguing today as
an example of his concept of the uniformity of ideas. As quoted above,
Vico declared that Homer was the Greek people, and that the two poems
were the treasure trove of Greek law. This, then, was the reason for Homer's
matchless faculty for heroic poetry – it was literally the culmination of
the entire Greek wisdom, mythology, religion and history. In effect, Vico
claimed that Homer himself was an imaginative universal, just as the Greeks
themselves had (unknowingly) used their gods as symbols of their own lives
and fears.

Vico was curious about the natural order of ideas which transcended
cultural boundaries and in the identical fashion in which the stages of
all cultures progress, not in the development or preservation of particular
nations.[10] He wrote in no uncertain terms that,

. . . ogni parola volgare dovette   . . . each vulgar word certainly
incominciare certamente da alcuno   must begin in every nation . . . [11]
d'una nazione . . .

Vico's argument regarding the uniformity of ideas between all peoples was
not strengthened by his refusal to admit any degree of cultural sharing,
either through cooperation or coercion, even among subgroups of the same
civilization.[12]

According to Vico language and early poetry must be rigorously and ruthlessly explored[13] in order to discover the early histories, the mental development and the way of thinking of these previous societies. He did not consider it a disappointment, but a confirmation of his approach that human ideas, laws and rights first developed from superstition, which he viewed as early attempts to pacify the natural world. These inborn ideas shared by all peoples at all times formed the basis of Vico's interpretation of *il diritto naturale delle nazioni* (natural law of the nations). These innermost ideas implied that man was not restricted to an instinctive or limited life.[14] Vico argued that a pattern of development was programmed into every human being before birth, but that it was done in such a way that man's will, as expressed by his creativity, was the very instrument of his individual cultural development. The idea of a *dizionario di voci mentale* which was common to all nations was closely tied to natural law. According to *La scienza nuova prima*:

E qui si pone fine a questo libro delle lingue con questa idea di un dizionario di voci, per così dire, mentali comune a tutte le nazioni, che, spiegandone l'idee uniformi circa le sostanze, che, dalle diverse modificazioni che le nazioni ebbero di pensare intorno alle stesse umane necessità o utilità comuni a tutte, riguardandole per diverse propietà, secondo la diversità de'loro siti, cieli e quindi nature e costumi, ne narri l'origini delle diverse lingue vocali, che tutte convengano in una lingua ideale comune.

We conclude our book on language with the idea of a dictionary of, so to speak, mental words, which is common to all nations. By expressing the uniform ideas of substance through which, by means of the various modifications, the nations thought about the same human necessities or utilities common to all, but looking at them through different properties according to their diversities of place, climate and, hence, nature and custom, this dictionary must narrate the origins of the different vocal languages, all of which share a common ideal language.[15]

Vico's natural law concerning the common nature of all nations and idea of the parallel development of human customs were coupled with his discussion of the development of man's mind and will in its various stages of development. Vico asserted that natural law, which he viewed as virtually identical to human customs, was born from tradition and that its roots were the eternal desires which all men shared. This was for Vico the foundation of the *umana mente*.[16]

4. *Sapienza volgare*

Sapienza volgare, also referred to as poetic wisdom, is fundamental to an understanding of Vico's thought.[1] He criticized the Epicureans and Stoics for, among other issues, having abandoned *sapienza volgare*. This warning against the dangers of an overly sophisticated approach is a recurrent theme in his work. Vico clearly stated that the first wisdom was not rational and abstract, but felt and *imagined*.[2] He struck another blow at the smugness and self-centredness of scholars since Plato, who believed that the first authors of language were themselves intellectuals. Vico emphatically denied that language was consciously created and that originally all words had been assigned meanings; on the contrary, he theorised that language had developed spontaneously, based on natural rather than prescribed significances for each word.[3] At the same time he regarded the development of language to be largely responsible for the development of the human mind, for it taught the mind to become more dextrous and swift in its workings.[4]

In the 1725 edition (Book II, 2) *sapienza volgare* was identified as the *senso comune* shared by all peoples and nations of the world. *Senso comune* was, therefore innate and had an educational aspect. Thus *sapienza vulgare* could be viewed as a refinement of the *senso comune* of a people. It follows that a concordance of national *senso comune* would be a record of the wisdom of all mankind. Book I of the 1725 edition demonstrated that the natural law of the nations arose along with their common customs. Vico delineated a progressive relationship among *senso comune*, *sapienza volgare*, the wisdom of all mankind and the natural law of the nations. What he described as the three *senso comune* of mankind – divine providence, an orderly family life based on civil religion and the burial of the dead – reinforced his three main precepts of religion, marriage and once again burial. In addition it reaffirmed his concept of the uniformity of ideas as demonstrated in *sapienza volgare*.[5]

Vico maintained that every group possessed its own version of *senso comune*, molded and fashioned by its own particular geographic, climatic and social characteristics.[6] As mentioned in Chapter 2, this second point has previously gone untold; even those who do mention *senso comune* in practice tend to equate it with the modern notion of divine providence, which is to say that it had perhaps only one or, more likely, no function at all.[7] At the same time most commentators would willingly acknowledge that the *dizionario di voci mentale*, although not *senso comune*, would certainly mainifest itself (as Vico clearly stated) in a multitude of forms. Further evidence in support of this new argument can be found again in Book I of the first edition, in which Vico wrote that common sense is the proof that men have free will –

thus the inference was clear that *senso comune* would vary radically culture by culture and age by age.[8]

Vico positively rejoiced in the poverty of articulate languages among the first nations:

Perché la povertá de' parlari fa
naturalmente gli uomini sublimi
nell'espressione, gravi nel
concepire, acuti nel
comprendere molto in brieve:
le quali sono le piú belle
virtú delle lingue.

For a poverty of words naturally
makes men sublime in expression,
serious in conception, and acute in
understanding much in brevity,
which are the supreme virtues of
language.

Queste sono le tre virtú più
rilevanti della favella
poetica: che innalzi e
ingrandisca le fantasie; sia in
brieve avvertita all'ultime
circostanze che diffiniscono le
cose; e trasporti le menti in
cose lontanissime e con diletto
le faccia come in un nastro vedere
ligate con acconcezza.

The three most important virtues of
poetic language are: that it should
heighten and expand [our powers of]
imagination; that it should give
brief expression to the ultimate
circumstances by which things are
defined; and that it should
transport the mind to the most
remote things and present them in
a captivating manner, as though
decked out in ribbons.[9]

This direct statement demonstrates that Vico held there to be a constructive and contemporary application of his *scienze nuove* (new sciences).[10]

There is a fragment of Vico's writing – unpublished in his lifetime and unincorporated into any of his major works – entitled (again by Nicolini) '*Idea d'una grammatica filosofica*' ('Idea of a Philosophical Grammar').[11] Throughout his writings Vico used linguistic terms – etymologicon, dictionary, grammar, mental language and mental words[12] – and used them as symbols relating to shared qualities of all peoples at all times. Indeed the *idea di un dizionario di voci mentali, comune a tutte le nazioni* (idea of a dictionary of mental words, common to all nations), the twelve concepts, most important as shared ideas, is particularly noteworthy in this respect as well, since Vico referred to them as *voci*, thus confirming his emphasis on philology and etymology as the means to comprehending past times.[13]

Vico saw old legends, fables, chronicles, laws and customs as having primitive history 'embalmed within', as Charles Vaughan (1854–1922) phrased it.[14] The recognition that poetry and myths were forms of cognition was the basis of Vico's assertion that the mind of past civilizations could be penetrated by means of imagination.[15] He distinguished between myths, the original stories passed on through oral tradition and fables as later editions of the same.[16] A fable must be a credible impossibility, inspire awe and possess an element of the supernatural.[17] At one point Vico declared that myths were true narration but that fables were false narration; he later realized that fables – if less useful in a study of the beginnings of a culture – were themselves true, faithful, contemporary reactions to past events.[18] Hence the origins of poetic characters were fundamental to Vico. They constituted the essence of fables;[19] for the fables themselves, which Vico called imaginative class concepts, *generi* or *universali fantastici* were the key to the mind of past civilizations.[20]

Breaking the code of the poetic characters was essential to Vico in order to discover the human necessities which they represented.[21] Although not inspiring reading, the clearest example of this principle is in Book V of the 1725 edition, in which he discussed in great detail the significance of the twelve major Greek and Roman gods. For example, he used Achilles as a paradigm, as the personification of virtue, Jove for idolatry and divination, Venus for civil beauty, Minerva for civil order and Mercury for commerce. Although these associations were not original to him, they were especially pertinent to his interest, for the gods were held to represent early human priorities, thoughts and emotions.[22] The modern study of pagan theology was essential,[23] according to Vico, to unlock these most fundamental human reactions to the world. Vico would not have been concerned if his divine and heroic characters and ages had never existed, the key point being for him that they existed in human memory as imaginative universals for human emotions, relationships and dramas. This theme is so dominant in his work that it seems almost redundant when Vico finally stated that poetic wisdom contained historical significance and that the first writers of both ancient and modern nations were poets.

It is now generally maintained that it is not possible to discuss the origins of language and mythology separately. Vico, as is well known, discussed them particularly in connection with the first and second stages of a civilisation. He considered original myths in the age of the gods to be narrations of real events or emotions, which were later misunderstood and altered to suit the times. Unfortunately, he gave no indication of how to distinguish the later, false myths from the original ones. But he did make several assumptions in regard to myth-making.[24] First, he wrote that primitive man had strong feelings and that they were dominated by their passions and bodily functions. Second, he argued that the thoughts of those early men were expressed in animistic forms – hence the myths themselves. Third, Vico presupposed a shared set of reactions for all men, which he sometimes discussed as *senso comune*, other times as a *l'ordine naturale d'idee* (natural order of ideas), and elsewhere as human necessities or utilities; Vico maintained that wildly divergent cultural groups experienced similar patterns of development in regard to the growth of their societies. Finally, he assumed, based on his other premises, that primitive myths must have a rough unity in terms of their subject matter.[25]

Vico gave credence to mythology in an era when it was generally dismissed as the product of superstitious minds and thus of no value to advanced societies, but the genuine appeal of mythology to Vico was that it was the product of primitive minds. The early modern and particularly eighteenth-century fascination with travellers's tales of supposedly primitive peoples was usually treated as light diversion, whereas Vico recognized that by sifting myths one could come to a better understanding of the science of history itself and thus one's own contemporary society.[26] For Vico the distinction between mythology and history was not useful. He realized, and it is now considered commonplace, that philology and comparative mythology could fill the gaps of philosophical speculation. He was not unaware of the similarity of linguistic and mythical concepts and their shared opposition to rational thought. Vico argued that language was man's prime instrument of thought, but he deemed that this initially reflected man's myth-making rather than his rational tendencies. Thus symbols were necessary for representations of any mental process. Hence we can better understand the importance Vico placed on the long passages at the end of *La scienza nuova prima* in which he discussed the various human emotions connected with particular classical gods and goddesses: for example, Hercules for heroism and Hermes for inventive intelligence.[27]

When Vico discussed the development of the human mind, it was imagination which separated the strong from the weak, the wise from the foolish,[28] the daring from the pedestrian. That the human mind takes a delight in uniformity, that early man judged the unknown by the known and that these same men created social institutions according to their own ideas, are all vital to an understanding of Vico's theory of the development of the human mind as allied to the development of culture.[29] Social change was not at all an alien aspect to Vico in the development of a society but was rather part of the social fabric.[30] The imagination was most acute when reasoning was absent: this was the primary reason Vico considered early societies to be rich in creative power. It was a great concern of his, in the 1730 edition, that later men were unable to understand early imagination.[31] Vico wrote that it was curiosity which led early man to explore his environment and that, understandably, they judged the unknown by the known.[32] He hypothesized that the first stage of poetry was divine because 'early man imagined the causes of the things they felt and wondered at to be gods' (*'i quali tutte le cose che superano la loro picciola capacitá dicono esser dèi'*).[33] He neatly contrasted this to the modern mentality, so civilised and detached 'that we can scarcely understand, still less imagine, how those first men thought who founded gentile humanity' (*'affatto immaginar no si può, come pensassero i primi uomini, che fondarono l'umanitá gentilesca'*).[34] The modern penchant for trivialities was for Vico not unrelated to the issue that 'the human mind takes delight in uniformity' (*La mente umana è naturalmente portata a dilettarsi dell' uniforme*).[35] For Vico, order and regularity were natural desires, but this was not necessarily meant complimentarily, for elsewhere he wrote that 'the weak desire laws, the powerful withhold them' (*I deboli vogliano le leggi, i potenti le ricuscano'*); further, and this would have been an insult coming from Vico, the weak interpret laws literally (*'Gli uomini di corte idee stimano diritto quanto si èspiegato con le parole.'*).[36] He discussed enthusiastically the capacity of the human mind:

La curiositá, propietá connaturale dell'uomo, figliuolo dell' ignoranza, che partorisce la scienza, all'aprireche fa della nostra mente la maraviglia, porta questo costume: straordinario effetto in natura, come cometa, parelio o stella in mezzodi, subito domanda che tal cosa voglia dire o significare.

Curiosity – that inborn property of man, daughter of ignorance and mother of knowledge – when wonder wakens our minds, has the habit, wherever it sees some extraordinary phenomenon of nature, a comet, for example, a sundog, or a midday star, of asking straightawway what it means.[37]

This is one of the few later mentions of curiosity by Vico, imagination having clearly replaced this faculty for him.

5. Religion and Society

It was not in Vico's discussions of sacred human history[1] that the close relationship between the Gentile religions and society, particularly the growth of societies, was established. Rather, it was in his analysis of social customs and religious rites. Religion, marriage and burial – the three human customs (*umani costumi*) which Vico asserted all civilisations shared, sometimes discussed as stages of utility, are mentioned time after time in all three versions of *La scienza nuova*.[2] Although religion seems an obvious way to examine the past in a Vichian manner (this approach is put forward by Berlin as rites of religion),[3] it is the one means (the other two being language and mythology), which Vico himself paid the least attention and which is the most difficult to use for the purposes of historical reconstruction. This is true in practical terms because the more advanced religious practices generally have, if anything, even less to do with earlier ones than epic poetry has to do with early fables and myths.

Rather than the Christian religion, Vico was concerned primarily with the mystical side of pagan religion – ritual and ceremonies. The religion Vico discussed at length was 'a "civil" phenomenon, profane and historical' (Karl Löwith, 1879–1973).[4] Vico would have agreed with Rousseau that the political religions of antiquity were false but useful, and that Christianity was true but socially useless, to the extent that Vico did not argue that religion, marriage or burials had a fundamental role in the shaping of societies in the later stages. By this point in the development of a society, he considered his three human customs to be rituals rather than spontaneous acts.[5]

It is now known that totemism, incest taboos, divination, as well as sacrifice, visions and ecstasy were natural parts of the religious life of many of the primitive peoples described in his first two stages – although only the third, divination, was cited by Vico. According to him, the authority structure of these early societies was reflected very clearly in their established rites of religion, and it was in weddings and burials that the overlapping of the customs of society and religion was seen most clearly. Marriage was essentially a socio-economic arrangement in Vico's scheme, but he asserted that its confirmation through a religious ceremony was necessary for the development of the correct civic spirit.[6] He believed marriage made men more dependent on religion, less warlike, more apt to be content with a monogamous relationship and also more industrious – all of which he considered essential for the proper development of an advancing society. Vico noted that mourning rituals were very often strictly delineated in early cultures, and throughout his writings there is the implicit warning that such rituals should be protected so that the world does not once again return to its former bestial state.[7]

With the issue of property Vico blended together the topics of pagan religion and society most smoothly, yet unfortunately without any critique of Locke, although Locke was mentioned in other contexts by Vico.[8] One of the reasons Vico cited for the need for proper burials was that it was necessary to establish correct boundaries for family owned property.[9] Likewise, Vico argued human society could not begin without marriage, which would have had a similar significance in terms of property. The two acts which he did delineate as being of key importance to property, in addition to burial, were the invention of writing and names for family relationships.[10]

At the base of Vico's model was social organization, mixed with what we would now term politics. Law, authority and succession were all viewed as part of the hierarchical structure. Vico dealt with age only in regard to hierarchy, not in relation to kinship. Nevertheless familial ties were the basis of his first stage of society. Vico stated this again and again, but never attempted to analyse or codify these various relationships. Kinship and social custom are now recognized as the key to any social structure; Vico saw the former as implicit, the latter as fundamental.[11] He mentioned the related areas: invention; subsistence; economic organization; social life beyond marriage and burial; fraternities; government in terms of laws, but not the development of government; art was mentioned rarely, if at all. However art is now recognised along with language, mythology and rites of religion as the fourth means of interpreting the past history of previous civilizations. Music was discussed occasionally along with early poetry.

Vico never dealt directly with the complexity of primitive societies, although most certainly he did see a given culture as much more than the sum of its traits. His references to social institutions were provided as a backdrop for these social issues, which in turn were to answer the issues of exactly what was human nature (primitive and modern) and thus history. He rarely commented on ethical issues; instead, the two areas he was deeply committed to were communication – signals, language, mythology, poetry and music – and rites of religion – prayer, magic and ritual. His credence in the unique quality of social institutions was underscored throughout *La scienza nuova*, but without reference to the complexity of social values.

6. Free Will

Vico declared that free will (*libero arbitrio*) was the 'artificer [creator] of the world'.[1] In *Vici vindiciae*, the 1729 response Vico wrote to an attack on *La scienza nuova prima*, published in Leipzig, Vico reinforced this revolutionary statement by affirming that philology, which elsewhere in his work is argued to be identical with history, depended on the free choice of man, language, customs, peace and war in history and proper philosophy.[2] In the 1725 edition Vico's second great principle, following divine providence and preceding the human will as the artificer of the world of nations, was vulgar or popular wisdom as the *senso comune* possessed by each people or nation. Thus the individual has *libero arbitrio*, in a social, never a theological sense, according to Vico. He compared it to *senso comune*, which demonstrated itself in wildly divergent manners for different societies.[3]

Vico's definition of the history of ideas was very closely linked to his ideas on custom;[4] he asserted, rather arrogantly, that his *scienza nuova* provided both a philosophy and a history of human customs[5] which could explain the varying customs and languages which came about because of differing geography, climate and culture.[6] According to Vico individual groups devised customs which were best suited to their own particular situation, thus he implied that there is some degree of free will. Yet this free will was always for groups, never individuals, and was not limitless even for societies, since social groups were bound by their environments – both natural and created.

It is the mental vocabulary of these social institutions which Vico so desperately wanted to translate.[7] The institution which he was most concerned with was not marriage, burial or rites of religion but the law. He discussed at some length[8] the importance of guarding the social institutions and, in particular, law. As Vico's primary aim was to study the social institutions created by man, it is hardly surprising that he clearly identified the power relationship, however variable, among decision-making, determination and the human mind. In the final edition little time is spent on *libero arbitrio*; rather, there is a shift to imagination.[9] Nonetheless it is necessary to note Vico's early discussion of free will, both in *La scienza nuova prima* and in *Il diritto universale*, in order to be able fully to comprehend his definition of imagination.[10]

7.   Formation of Society: The Taming of Primitive Man

In 1744 Vico stated boldly that the role of philosophy was to raise and direct man, the purpose of poetry to tame the ferocity of the vulgar, the purpose of legislation to turn man to good use in society and the purpose of religion to subdue the savages.[1] These forces were all designed to build efficient societies in which the individual would gain satisfaction from fulfilling his particular place in society. Vico's scheme of social engineering included the taming of man, domestic education, the end of roaming and the nomadic lifestyle, the consequent development of home and homelife and the eventual development of familial authority into civic authority. All of these were underlying themes of Vico's theoretical work on societies,[2] including even the human tendency which Vico's work completely depended upon, that 'men are naturally impelled to preserve the memories of the laws and institutions that bind them to society' ('*Gli uomini sono naturalmente portati a conservar le memorie delle leggi e degli ordini che gli tengono dentro le loro societá'*).[3] This was itself a recognition that there are forces within man that are not of his own making.

But the key to this taming of primitive man lay not in Vico's lengthy discussions of human needs and utilities but in the concept of shame, modesty (*pudore*), the prime motivating and civilizing force, moving men closer to the traditions of monogamy, a sense of family and beliefs regarding gods and religion.[4] *Pudore* was also critical for Vico as it implied that the sense of conscience was an innate quality. Hence it was for Vico not a positive but a negative force – shame – which was the basic power, propelling man out of his original bestial state.

In the midst of this discussion of the taming of primitive man Vico maintained that poetry, as well as philosophy, played an educative role, teaching the vulgar to act piously.[5] As has been mentioned, he similarly maintained that marriage made man more amenable to group pressure, more peaceful, monogamous and industrious.[6] Thus Vico was not merely concerned with the creative aspects of poetry, or with marriage as a religious rite. He was also determined to discover the link between imagination and the creation of a social group, the taming of primitive man into a fully functioning member of society. The reason that fables were intended to frighten and excite the people was precisely so they would enter into and manipulate the minds of their listeners. These fables had to be suited to popular understanding, otherwise they would not achieve their purpose. But there is a problem here. Elsewhere he discussed the importance of free will in history – it is one of the contentions of this work that Vico did not believe the cycles to be predetermined – nevertheless, Vico assigned to poetry an almost determinist role in early society which is closely aligned to the traditional Catholic interpretation of Vico's divine providence.

Although there is no uncertainty that Vico was most concerned with the early stages of early society, he never adopted a wistful attitude towards this period. Nor is there any trace in his work of the *bon savage* (noble savage) mourned and all but venerated by Rousseau. Of greater worth than the dissection of Vico's stages of a civilisation is an examination of his usage and interpretation of the past, which was not pre-Romantic; rather Vico freely criticised the same peoples and societies which he acknowledged left the richest cultural heritages.

Very often Vico contrasted the evils of the beginnings of a social group with the mirror image failings of the final stages of a society – the primitive man with the overly refined one. As already mentioned in the section on *De nostri temporis studiorum ratione*,[7] Vico struck out against conventional scholars[8] as he argued that the imagination of the young was dulled rather than encouraged by contemporary educational practices. Vico had little time for civilised minds with their abstractions and refinements,[9] but neither did he delight wholeheartedly in primitive attempts at comprehending the world. Yet the imagination of these early stages is certainly for Vico the tool which created these social institutions as well as the code-breaker of past civilisations. One peripheral reason for this is that Vico did not expect imagination to manifest itself in the form of epic poetry in the very first years of an incipient social group, although imagination was demonstrably at work in his first age, that of gods.

Some of the same tension in Vico is also present in the work of Hobbes, whom Vico much abused. In Hobbes one is never sure how primitive man ever evolved out of his feral state. In Vico there is a similar ambivalence towards primitive man, indeed it was men not so far removed from the same *'primi uomini, stupidi, insensati ed orribili bestioni'* ('first men, stupid, insensate, horrible beasts')[10] who had the *robustissima fantasia* ('extremely robust imagination') which so fascinated him. Yet Vico provided transitions and progress by means of his discussion of the gradual development of a culture. But it must be noted that he wasted little time on even stating how the shift occurred from one stage or level to the next. Hence the problem for those who would like to apply Vico's principles of *'la storia ideale eterna comune a tutte le nazioni'* is that they are given no instructions and set no boundaries regarding how exactly to do so.

Vico listed numerous groupings of three – for example, three ages of poets before Homer, not to mention three kinds of natures, customs, natural law, governments, languages, characters, jurisprudence, authority, reason, and judgements. Vico's groupings of three were done for the sake of convenience (and out of personal preference) but he never forgot that they encompassed innumerable changes and developments.[11] The most interesting group of three was not original to Vico – that is the development of the human mind from senses to imagination and finally to rational thought, which clearly echoes the division of mental abilities and society in Plato's *Republic*.

Authority – a key term in Vico's vocabulary – was certainly not a later development of a society. According to Vico:

... che tra'l debole e'l forte
non vi è ugualità di ragione;
perchè non mai gli huomini
patteggiarono co'leoni; nè le
agnelle e i lupi ebbero mai
uniformità di voleri: ...

... there is no equality of right
between the weak and the strong,
for men have never made pacts with
lions nor have lambs and wolves
ever shared any uniformity of
desires.[12]

The concept of basic rights for all was a later recognition,[13] for Vico believed that it took a developed and cultured mind to recognise and enforce common privileges for all classes.[14] Many diverse and unusual types of authority figures were referred to by Vico – religion and poetry had pride of place as the twin authorities of his science, but he also gave specific examples such as fathers, kings, philosophers, lawgivers and scholars, to name just a few. In his discussion of natural law, Vico's civic hierarchy is clearly delineated: (1) the natural law of nations has only to do with civil authorities, (2) these civil authorities should be revered as sacred persons who recognize no superior other than God, (3) these rulers have right of life or death over subjects.[15] Although Vico recognized that there was a multitude of varieties of societies and governments, this in no way indicated that he approved of a flexible approach within any single society. Vico then blended the two diverse topics together by claiming 'authority as the free use of will'.[16] Natural law was of supreme interest to Vico, since it was for him identical with custom. The authority structure and community life fashioned human nature, and not *vice versa*. According to Vico, people are directly affected by their social and physical worlds and heritage. Free will of the community determines the social and religious mores of that society. Therefore his more unusual discussion of natural law[17] traced the authority of human nature to the authority of natural law, which was for Vico human customs and thus culture.[18]

8. *Setti di tempi*

Vico employed the terms *modus* (mode, measure), *forma* (form) and *genus* (class, kind), interchangeably in *De antiquissima italorum sapientia* to refer to the manner or style of a particular society at a particular time.[1] This was at the very heart of Vico's search to identify the manner in which previous societies lived and died. This again supports the view that the *senso comune* of each group is manifested in a somewhat different form, although composed of the same basic elements.

Although it is Vico's concept of *una storia ideal eterna* that is most often cited by those making passing reference to his work, this term is very often misleading, particularly when discussed with regard to the cycles presented in the final edition. The notion of an ideal, eternal history only makes sense in the context of *senso comune*, and it can only be investigated by means of imagination.[2] In the first edition of Vico's *magnum opus* he did not discuss the three ages at all. It is in Book V of the final edition that one finds his well-known list of threes – characters, jurisprudence, *autorità*,

reason, judgements and *setti di tempi* (divisions of time), to name a few
– these stages of development do not vary significantly among the diverse
topics, rather they are similar in pattern to the growth of a language which
corresponded to the appropriate stage of a civilisation.[3]

Without a doubt, Vico was entranced by the uniformity in the development
and decline of nations – he named these stages the divine, heroic,[4] and
human ages.[5] He desired to discover the mentality of these early peoples
via their particular customs. There was a clear link between the origins
of language and his principle of the development of nations. Vico stated
without qualification that 'the world of human nations has certainly been
made by man' ('*che questo mondo civile egli certamente è stato fatto dagli
uomini*'),[6] and for Vico man made society.

A common interpretation of the 1744 edition presents these periods,
these divisions of time, as of more importance than the cultures which
flourished within them and which of course overlapped these arbitrary
boundaries. Quite the contrary, in his attempt to penetrate the nature of
nations, Vico always leads us back to the common mind possessed by
all peoples.[7] He cautioned against the hope that language, for example,
would answer all questions about preliterate mentalities, for if we cannot
make contact with feral children (a common preoccupation, even before the
early nineteenth-century Kaspar Hauser *cause célèbre*) how do we ever hope
to understand those who lived so many centuries before, in such a diverse
manner? Thus Vico employed language not because he considered it capable
of answering all of our questions regarding early societies, but rather because
it offered a possible avenue to the past. It was because the mind of the nations
diminished – in terms of collective imagination, Vico's main preoccupation
– with the development of literacy[8] that he was drawn to early societies.

Una cronologia ragionata ('a rational chronology'), *un gran mostro di
cronologia*' ('a great monster of chronology'; a play on words by Vico,
mostro as opposed to *mostra*, demonstration),[9] was essential to Vico's
new study of history.[10] Clearly the cycles are an essential part of his
historical critique, but all too often this has been given undue importance
by commentators. Almost always he focused on the first stage of a society to
the exclusion of later developments; yet, by reiterating that there were at least
three principal phases, Vico regularly reminded his readers that he was not
unaware of the importance of later developments. Vico was less engaging,
certainly less original on the decline of a society. It was his aim to seek out
the roots, the origins of later, more sophisticated developments.[11]

In the 1725 edition Vico discussed at length another example of *la boria
de' dotti* – of assuming some sort of understanding of societies of the past
which were not separated by a long time period.[12] For Vico an anachronism

was best defined as 'perverted time'.[13] He criticized traditional historians for assuming that there were no events of importance in the earliest times. These common errors of chronology received scathing comments from him, particularly because he believed that from his time forward his science of imagination could circumvent them. Vico's stress on the notion of an anachronism owed much to his reading of Renaissance writers on philology and history, notably Valla and Bodin.[14]

As mentioned in Chapter 2, Vico's 1725 definition of natural law was very broad indeed including as it did the origins of religions, languages, customs, positive laws, societies, government, types of ownership, occupations, orders, authorities, judiciaries, penalties, wars, peace, surrender, slavery and alliances.[15] According to Vico, natural law was composed of all the tangible aspects of human society, and study of these institutions was the correct means to gain insight concerning a past society. In addition to the issue that Vico believed that this *il diritto natural delle genti* (natural law of the nations) was composed of virtually all elements of society, he also argued in a circuitous fashion that it proved the validity of what may be viewed as the most controversial of all the elements, that is, the Christian religion[16] – a point which would have made rather more sense if it had referred to the fundamental role of early pagan religions in the development of a society. Even though Vico reiterated numerous times that his interpretation of natural law was Christian and worked out by divine providence, nevertheless there is nothing of the sacred or religious in his discussion of natural law. Vico's natural law dealt only with the workings of Gentile, non-Judeo-Christian history, which was the only history he ever discussed in any detail. Vico believed that natural law, divine providence and his ideal, eternal history were the principles on which human history was worked out, but his main focus was not on these abstract concepts as much as on human institutions, the study of which was the only way to penetrate human history and consequently to comprehend natural law.

In the 1744 edition natural law was listed simply as the sixth of seven aspects of his new science. It followed divine providence, a philosophy of authority, a history of human ideas, a method of philosophical criticism (based on the history of human ideas); an ideal eternal history preceded the principles of universal history.[17] For Vico (unlike Plato) the order of knowing was the same as the order of human institutions, by which he meant that one could only truly comprehend that which one has made; thus he stressed once again the notion of natural and gradual growth and development. He expressed natural law in slightly different terms:

| Il diritto natural delle genti | The natural law of the gentes is |

è uscito coi costumi delle	coeval with the customs of the
nazioni, tra loro conformi in	nations, conforming one with
un senso comune umano, senza	another in virtue of a common
alcuna riflessione senza prender	human sense, without any reflection
esemplo l'una dall'altra.	and without one nation following
	the example of another.[18]

Again Vico stressed the ability of separate nations to develop in a parallel fashion without any contact having taken place between them.[19]

In *Vici vindiciae* Vico claimed natural law as central to his science, believing that the social side of man's nature was inextricably linked with, if not identical to, natural law.[20] By means of this connection of nature and custom, Vico then spoke of natural customs (*i naturali costumi*) thus destroying the long opposition between nature (*physis*) and custom (*nomos*). Hence 'reasonable customs' (*consuetudine ragionevole*) were natural as they arose from man's own nature, and following them was both natural and pleasant.[21] For Vico these 'natural customs' were very similar to his own 'principles of humanity', often discussed as 'these universal and eternal principles . . . on which all nations are founded and still preserve themselves' ('*i princípi unviersale ed eterni, quali devon essere d'ogni scienza, sopra i quali tutte [le nazioni] sursero e tutte vi si conservano in nazioni*') and 'those institutions which all men agree and have always agreed' ('*quali cose hanno con perpetuitá convenuto tutti gli uomini*').[22] Vico asserted that he had solved the long argument between customary and natural law by blurring the distinction itself. For Vico the question 'is man naturally sociable?' would have carried the same implications as 'does natural law exist?'. Thus, he concluded, society was itself natural because of law, which was based on some customs, which were themselves derived from man's needs and utilities, which were basic to human nature.

Vico's definition of natural law may be viewed most appropriately as an elaboration of his ideas regarding the development of the human mind and of nations. According to Vico, every person, every social group, shared certain common ideas regarding their position in the natural world, and these ideas were arrived at through common sense. Human needs and utilities were the two sources (*fonti*) of natural law. Common sense functioned as the foundation, since it provided the means of defining what was certain (and thus necessary and useful) in natural law.[23] But this is not to say that Vico adopted a utilitarian approach to natural law: because of his religious beliefs, he adhered to an eternal order of ideas, which could not have arisen from the natural body alone. Vico refuted the utilitarian theories of the development of law; he was repelled by the materialism and relativism of an approach which stressed either the demands of the weak or the dominance of the strong – by means of his conception of the law as an eternal truth inherent in the nature of man.

By means of his definition of natural law Vico maintained that the popular notion of law could be analysed and the social and civil experiences of past societies could be *riconosciuto* (known again, recognised). Vico viewed the natural law as a 'jurisprudence of mankind' (*'una giurisprudenza del genere umano'*).[24] He deemed one of its greatest strengths to be that it offered a method, 'a new art of criticism' by which to analyse the barbaric stages of past civilisations. It provided the philosophical groundwork needed to explain the *certa mente comune*,[25] which was quite distinct from any form of antiquarianism. Without a doubt, Vico incorporated his interpretation of natural law, which dealt with customs, shared cultural characteristics and *senso comune*, into his historical and philosophical scheme, as it offered him another avenue by which to analyse early civilizations.

9. New Critical Art

Vico felt compelled to discover and devise a new critical art, his new science,[1] in which even physical abnormalities, such as giants[2] were of importance exactly because they demonstrated the truth – by which he signified the fundamental role, usefulness, veracity on their own terms, and dependability – of the fables. He considered it necessary to elucidate rules which would enable one to discern the truth of all Gentile history, and in this manner facts, laws and nature would all be interpreted in light of one another.[3] 'Vulgar traditions given in verse must be true' Vico wrote in 1744, 'because this form is so old' (*'Onde di tal spezie di verso bisogna che sieno vere quelle volgari tradizioni'*).[4] By 'true' he did not mean correct in all particulars, but rather that, far from detracting from the usefulness of these myths, the inaccuracies themselves reflected the attitudes and mores of subsequent eras.[5] Thus every stage or layer was a true representation of a particular culture. Vico's method was not an all-forgiving look at the past, rather it was intended as constructive criticism.

In the first edition Vico clearly stated the two practical aims of his *scienza nuova*: the first was a new art of criticism 'by which to discern what is true in obscure and fabulous history' (*'una nuova Arte Critica, che ne serva di Fiaccola da diftinguere il vero nella Storia Ofcura, e Favolofa'*) and the second an 'art of diagnosis, recognition of the indubitable signs of the state of the nation' (*'oltre quefta l'altra Pratica è un' Arte come Diagnoftica . . . di conoscere i fegni indubitati dello Stato delle Nazioni'*).[6] He maintained that his rules of interpretation were applicable even to new laws and fact.[7] It was truly a new means of discovering the past.[8] In 1725 he stated that his new art of criticism revealed the whole of Gentile theology, which included new ways of thinking about pagan religions as a means to discover the culture of the relevant societies.[9] When in 1730 Vico discussed the two branches of his study, *poetica metafisica* (metaphysical poetry) and *scienze specolative* (speculative sciences)[10] his goal was to combine the two to form a *scienze poetiche* (poetic sciences).[11]

The term *degnità* (axiom), used in *Il diritto universale*, does not appear in the first edition of *La scienza nuova*; only in 1730[12] and then in 1744 did he use this already archaic term.[13] It was in the second edition, and in his first *degnità*, that he exhorted the reader to use his own imagination to comprehend early man. The use of this term is crucial, for it indicated the coming together of Vico's thought on the critical method. In this same edition he asserted that 'with our intellects we can amend the errors of our memories' (*'pruove le quali soddisfacciano i nostri intelletti, sono ammende che si fanno agli errori delle nostre memorie'*).[14] This was in effect a clarion call to apply the critical method to fables, customs, religious rites, laws and all other such remnants of previous cultures.

Vico maintained that the reason that such a science was lacking was that no one had previously addressed the history and the philosophy of humanity together.[15] In '*La pratica*' he wrote boldly that his was a science of contemplation from which could be derived a study of the development and decline of nations.[16] Philosophers and philologians (which for Vico included poets, historians, orators and grammarians)[17] should together examine the wisdom of ancient Gentiles by means of a study of all the diverse records left by primitive man.[18] Vico's goal was nothing less than a history of human ideas.[19] He well understood that this could not be accomplished easily; on the contrary, he considered the best way to understand such a history was to investigate the 'natural order of ideas . . . religions, laws, languages, marriages, names, arms and governments proper to them' (*'Ordine naturale d'idee dintorno al diritto delle nazioni per le loro propie religioni, religioni, leggi, lingue, nozze, nomi, armi, e governi'*).[20]

Book II of the 1730 and 1744 editions once again elucidated the common principles, in this case of articulate language, which Vico believed was his particular task. He postulated that one could use any language (in his case, Latin) to discover the true roots not only of that particular language but of any tongue. Language, which he called 'a mighty witness of the ancient customs of the peoples', was for him the natural key to understanding past civilisations, and he believed the origin of language was to be found in early poetry and mythology. Vico maintained that the genius for poetry was a gift from heaven; at the same time he also considered primitive man, with his lively imagination, to be particularly well suited for this form of self-expression. Vico defined fables as a way of thinking for an entire group. He was concerned not only with the actual poetic universals formed by primitive groups as a recreation of their own lives and histories, but in addition he considered at least as important the instinct which led to the preservation of these laws and institutions.[21]

The uniformity of ideas among all nations at any stage of development was often discussed as the *certa mente umane* or *certa mente comune*.[22] It is hardly surprising that Vico considered the immediate need to be a method to discover this common mind possessed by all peoples at all times. The term *sapienza* (wisdom, learning and knowledge) was very closely related to Vico's statement of the need for a scientific, critical, trained and skilled grasp of the customs of the nations, by which he meant the blending of early wisdom as demonstrated by social conventions and customary, practical theology and morality. It is this recondite, hidden, obscure, abstruse and above all profound wisdom which was his goal.

Vico clearly recognised the difficulties inherent in devising a means to decipher the ancient, primitive languages and especially their developing vocabularies, which, he was persuaded, represented the development of their own ideas. Much the same is his later admission regarding the dilemma of historical reconstruction.[23] In 1725 he wrote of the difficulty of reconstructing the world of ancient Rome, for the information an historian possesses of a past culture is so fragmentary.[24] For this reason he asserted that it was essential to have a philosophy of history in order to reconstruct past societies and mentalities.

Vico described an ideal, eternal pattern of history which supposedly dictated the development and decline of nations. He examined the human condition in terms of a cyclical theory of history in which civilisations progressed through three stages – from the age of gods, to heroes, and finally men. The key role is played by classical and popular mythology in his attempt to unravel the pattern of history within this speculative framework. Vico regarded the science of myth to be his greatest achievement. Arguably the single most exciting passage from the final edition is the following:

. . . vi si vaglia dal falso il vero in tutto ciò che per lungo tratto di secoli ce ne hanno custodito le volgari tradizioni, le quali, perocché sonosi per sí lunga etá e da intieri popoli custodite, per	Truth is sifted from falsehood in everything that has been preserved for us through long centuries by those vulgar traditions which, since they have been preserved for so long a time and by entire

una degnitá sopraposta debbon
avere avuto un pubblico
fondamento di vero.

peoples, must have had a public
ground of truth.

. . . i grandi frantumi dell'antichitá,
inutili finor alla scienza perché
erano giaciuti squallidi, tronchi e
slogati, arrecano de' grandi lumi,
tersi, composti ed allogati ne'
luoghi loro.

The great fragments of antiquity,
hitherto useless to science because
they lay begrimed, broken, and
scattered, shed great light when
cleaned, pieced together, and
restored.

. . . sopra tutte queste cose, come
loro necessarie cagioni, vi
reggono tutti gli effetti i quali
ci narra la storia certa.

To all these institutions, as to
their necessary causes, are traced
all the effects narrated by certain
history.[25]

Vico was convinced that he had found in history what he had earlier
searched for in jurisprudence, and although he believed he had isolated
eternal truths (for example, '*la storia ideale eterna*'), he recognised that
his greatest achievement was the creation of a method which transcended
the mere acquisition of facts.

4 Language, Historical Reconstruction and the Development of Society

1. Introduction

Vico was not concerned with poetry as either art or literature; rather it was man's innate desire to create representations of the past which was his primary consideration. Neither did poetry in Vico's thought have anything to do with the products of sensitive, perceptive or refined imaginations, nor was it the work of individuals.[1] Vico showed no interest in the great artistic traditions which often accompany extravagant and decadent civilizations. His emphasis was always on the spontaneous creations of social groups rather than well-planned individual works of great genius. He did not discuss *la questione della lingua* (the question of language), which runs throughout the history of Italian culture and concerns the variety of language most appropriate for literary use.[2] For Vico early poetry was produced by necessity, not for pleasure. Croce was absolutely correct; it is necessary to understand Vico's idea of poetry if one wishes to comprehend *La scienza nuova*.[3]

Even in his theoretical discussions, imagination was for Vico the creative ability of a social group, the spirit of their time, and the ability of later groups to reconstruct these lost worlds. Vico showed no particular interest in the role of the individual or in the nature of man, as did Bacon or Hume. His recognition that language was a representation of a society never led him to the study of dialects or the varieties of language within a culture – although it would have been a natural extrapolation from his own ideas. Vico's mind was set on the abstract concepts of language, society and history, and he spared no time on the subgroups which might in many cases have led him much more quickly to his ultimate goal.

The three most general means of examining language have to do with its origins, its structure and the relation between language and reality.[4] In Vico, for the first time, one finds reference to all three. Donald Kelley links Vico with his sources:

. . . the grand tradition of the *ars poetica* going back to Boccacio, [on] which Vico draws: poets as first philosophers, poetry as the highest wisdom, origins of language, and other highly conventional themes – on which, of course, Vico performs his characteristically virtuoso variations.[5]

The trend towards formality in both speech and writing in the sixteenth and seventeenth centuries, and the increasing dominance of the Tuscan language led to a growing awareness in the Italian states of language as an active force in society.[6] Neither Descartes nor the Port Royal Grammarians had shown any interest in the origins of language, yet this peculiarly eighteenth-century fascination was one of the cornerstones of Vico's thought, even in his first writings at the end of the seventeenth century, no doubt because of his grounding in Renaissance thought.

Vico always turned to the earliest stages of a language for information regarding the development of societies. Although he devoted some lines in each of the three editions of his last work to the development of the parts of speech (interjections, followed by pronouns, particles, then nouns and finally verbs, according to Vico), this discussion is not critical to his most profound comments on language.[7] It is his arguments regarding language and the progression of cultures which are of the most lasting value, for history functioned as the vehicle by which further knowledge regarding the human condition and societies could be obtained.

Vico's views on language predated those of James Burnett, Lord Monboddo (1714–99), *Origin and Progress of Language* (1773–92), yet Vico's views on the subject are often interpreted in the light of Monboddo's later, better-known theories: that language was a human invention and that it was the natural product of the biological evolution of the species.[8] Vico's basic view of language was that it was a product of incipient social units and that among the first peoples the urge to express their feelings was innate. Hence, according to Vico, language was neither dependent on a fully functioning social group, nor did it have to be taught in a formal sense. Monboddo, in contrast, was to assert that the Greek language was the invention of a literate people. In addition Monboddo dismissed the examples of primitive peoples with advanced forms of a language as the result of language mixture and corruption, a possibility that Vico violently opposed.[9]

Vico would have been happier with Rousseau's view that the creation of a human society presupposed the existence of a language.[10] But, more precisely, for Vico the development of language and society were aspects of a single process, of interacting human faculties. Monboddo wrote 'Monkeys are in all ways human except they lack the ability to speak . . . ' and ' . . . if there were nothing else to convince me that the Ourang Outang belongs to our species, his using sticks as a weapon would be alone sufficient'.[11] Vico never subscribed to this later view that the linguistic development and abilities of human beings could in any way be described simply as a higher form of animal communication.

It must be stressed that when Vico spoke of language, he did not mean simply the articulate versions of the same, but all forms of human communication: gesture, exclamations, all the stages of the development of a language and even heraldry. In this sense Vico's view of what constituted language is much closer to that now ascribed to Monboddo than Vico might have cared to admit (had he been given the opportunity). Certainly the late eighteenth-century debate of the *ferini* (beasts) versus the *anti-ferini*, in which the supporters of Vico were labelled the *ferini*, recognised this tendency in his work.[12]

2. Vico and Early Language

Few authors when commenting on Vico neglect to mention his fascination with language, mythology and rites of religion – if only in connection with the bestial stages of his so-called cycles of history. Yet often these three topics are mentioned in such a way as to reinforce the accepted view of Vico as an eighteenth-century polymath. Indeed Vico's collected works fill eight volumes[1] ranging from jurisprudence to poetry, treatises on language (Latin in particular) and education, eulogies, orations, speeches written for prominent civic leaders to read at festivals and other occasions, inscriptions and dedications – as well as the theoretical works already discussed. It is always a somewhat dangerous enterprise to force eighteenth-century writings into twentieth-century categories; yet even judging by the standards of his time, Vico's works transcended the normal boundaries, not only in terms of subject matter but in their conclusions. Indeed, one would expect nothing less from the man who is considered the first modern philosopher of history, the father of the social sciences and the precursor of Hegel – to name but a few of the labels which have been attached to Vico. Without wishing to belittle the extensive range of subjects with which Vico dealt, but in order to have a better understanding of Vico's contribution to them, it is particularly significant to examine exactly what intrigued Vico about the pivotal topic of language and how it formed the basis of his new method of historical investigation.

Since Vico wanted to discover the first roots of society, he was forced to surrender the implicit assumption that it is only European classics we must study.[2] His research was thus centred on subjects excluded, for whatever reason, from the texts. Vico was intrigued by areas which would now be classified as ancient history or anthropology rather than history in the traditional sense. His theoretical studies dealt primarily with non-literate societies even though, as has been noted, his most numerous specific examples were drawn from a civilisation which was exceptionally well documented. Throughout *La scienza nuova* there are scattered references to North American Indians, Chinese, ancient Egyptians and Greeks, to name a few, but the bulk of his examples were drawn from classical Rome.[3] However Vico generally gave short shrift to the historical periods for which he might have gained documentation. There is no indication that he sought out what would now be termed archives or even the private libraries of Naples in his search for knowledge of the past.[4] This practice was certainly not unusual for the time. But Vico was so advanced in many of his conceptual methods that one would not have been surprised if he had anticipated this type of primary research. Yet as his own rather unexceptional historical writings reveal, Vico was never at his best when it came to the practical application of his theories.

Without a doubt, language was for Vico the natural key to understanding past civilisations and he considered that the origin of language to be found in early poetry and mythology.[5] He wrote that the 'genius for poetry was a gift from heaven' (*'poëticus instinctus Dei. Opt. Max. donum est'*),[6] but he also believed that primitive man, with his lively imagination, was particularly well suited for this form of self-expression. Croce expressed it in this manner:

. . . perché l'uomo rozzo e di debole cervallo, non potendo soddisfare il bisogno che prova del generale e dell'universale, foggia a sostituzione i generi fantastici, gli universale o caratteri poetici, . . .	Since uncivilized man is of low brain power and cannot satisfy the thirst he feels for the general and the universal, he fills their place by inventing imaginative genera, poetical universals or characters.[7]

Vico viewed historical reconstruction through philology as the most straightforward means of analysing the past, and, as he constantly reiterated, Latin was particularly well suited for this type of study. In *'Nova scientia tentatur'* Vico attempted to reduce philology to scientific principles, and he claimed *La scienza nuova* to be the first union of philosophy and philology.[8]

Vico's emphasis was on the origin of language rather than the distinction between languages. Geography and climate answered for him the question as to why there are so many languages.[9] The independent origin of similar grammatical processes would not have surprised Vico in the least, because he presumed that there were cycles in the development of language.[10] The issue of the scope for originality within a predetermined cycle of development of language in the Vichian scheme is thus just as relevant as that within a predictable pattern of development for societies. He wrote that as primitive man moved towards life in a village and then a city, his vocabulary, syntax and general command of his language developed accordingly.

But although Vico was engaged with language in all its aspects – philology, hieroglyphics and demotic language, to mention just a few,[11] his attention was on early language, both in spoken and written forms. Mythology was for him a way to regain early speech. Vico himself slipped relatively easily from one idiom (for example, scholastic, neo-Platonic and literary) to another in his writings.[12] In the same way he recognised that people have always used different codes on different occasions and in different languages, depending on audience, setting and topic.[13] The necessity of historians to study the oral form of the language through the written and investigate social groups via the records of other social groups was discussed innumerable times by Vico.[14] This type of study was of necessity complicated by the changes in form of the vernacular, which were accompanied by changes in the form of the language as a whole because of the need or desire for a standard, correct form of the vernacular.[15] Vico prized codified versions of early law as remnants of the early societies in which they developed.[16] Yet the standardisation of the language also marked the end of his poetic era, and after this point the language was less compelling for Vico in terms of its potential concerning the cultural study of societies' earliest phases.

One of Vico's greatest insights was that myths and legends carried significant subtexts. They were forms – even if corrupted ones – of the past history of a society. Yet hand-in-hand with this theory goes one of his work's shortcomings: Vico closed his eyes to the literal meanings of at least some of the classical legends he cited.[17] His discussion of myths and legends expressed a certain ambivalence, for he equated the rise of idolatry and divination with the establishment of the birth of the fables. Vico made no effort to hide his contempt for the pagan religions of which the fables were the focus, even though he considered the classical gods to be primary sources of human emotions, be they fear, jealousy, love or any of the passions.[18] Vico was greatly concerned with the very pagan religions he despised – for he maintained that their funeral rites, marriages and other religious ceremonies were inextricably linked to the beliefs and values of those societies.

Vico seemingly found no negative aspects of language at any stage, while at the same time he was often scathing in his criticism of the early men who spoke these first languages. Language had for Vico a curiously amoral quality. For him language could only reflect the stage of development of the society in which it was present, thus not affecting early societies directly; its role in any social conditioning was thus muted. For this reason he never regarded language as a social force for good or evil, discounting any sort of moral bias or basis for it whatsoever. This is also the reason that he discussed language at such length, because he regarded it as another way, perhaps the best way, to gain access to past societies.

Vico used the term language (*lingua* in both Italian and Latin) to denote all forms of human communication – not just articulate forms of speech and writings. In addition he used the word in four other major senses: firstly, as an instrument of thought; secondly, as synonymous with early fables, myths and poetry; thirdly, as parallel in its growth to that of the relevant society, while at the same time as an educational tool and social force in its own right; and finally, as perhaps the most important tool in historical reconstruction and thus a means, as well as a source, of historical knowledge. The term 'philology' (*philologia* [Latin], *filologia* [Italian]) was often used by Vico as a synonym for history, which for him comprised the history of civilisations, cultural development and social development. Language was for Vico the key to identifying these essential stages of a society.

Thus the single word *lingua* was used by Vico to convey a multiplicity of ideas, both commonplace and original. There was some discussion of the structural development of language. He accepted that language had formal characteristics (such as an inflexible pattern of development of the parts of speech within any given language). At the same time he also viewed language as man's most fundamental attribute, mainly because he maintained that language was the means of discovering human creations, particularly social institutions. He regarded language as a human invention because in his opinion language only developed in social groups, not spontaneously within each individual as a natural part of his particular physical development.

Vico's blatant lack of interest in later, more developed stages of a society is nowhere more apparent than in his discussion of language. Early societies were the root, not the fruit, of Vico's thought. He was entirely focused on development in antiquity, on beginnings. There is no evidence in his work of any appreciation of the creative arts or even the poetry of his own time. According to Vico modern man had lost the protective covering of accepted myth structures. This loss of myth-consciousness was to him both essential to the development of a society and marked the end of imaginative creation, Vico's most compelling interest.[19] When Vico discussed language, it was almost always the language of primitive peoples, never of partially developed societies or those of his own time. His concern was concentrated almost exclusively on the early phases of a civilisation, not even on the transitions from it, much less on the apex or decline of a state. Philosophy was the antithesis of poetry, according to Vico, and once philosophy gained dominance in a society, poetry as a living art form withered away.[20] This early poetry had to do with basic human needs and the religion and laws of the first peoples.[21] Although poetic activity enjoyed maximum autonomy in his final, non-poetic period, this phase was of little concern to Vico, because the poetic forms which preoccupied him had nothing to do with the free and creative urges of gifted individuals. Much more importantly, Vico saw early poetry as a source of cultural, ethnological information.[22] Vico's concern with societies is central. His approach arose initially through an examination of jurisprudence. Vico maintained that the authority of poetry resided in common humanity, in its development and historical conditions and that it was not the result of change or the abstract consideration of principles.

As man's poetic qualities – cruelty, violent passions, blind self-interest – would by themselves tear society apart, Vico argued that it was divine providence which saw to their utilisation and historical efficacy.[23] This abstract, but for Vico very real, force within his philosophy of history was not unlike the secular versions presented later by Hegel as the 'cunning of reason' and by Adam Smith as 'the invisible hand'. The real history in Vico's writings was presented and discussed in the form of unintended consequences. Antiquity demonstrated for Vico the perpetuity of sacred history; because of the critical role of divine providence, the most important histories for him were secular as they were the points of entry into the changes and successions of the past. It was the elements of pagan history which concerned him – most particularly, law, which he believed shed light on both human institutions and words.

As discussed in Chapter 2, Vico wisely pinpointed two causes of the ignorance surrounding the origin of poetry – thinking that the language of poetry was peculiar to poets not the people at large, and the belief that it was the poets who founded the first religions (*De constantia iurisprudentis*, XII). Just as Vico considered myths to be the product of the mind of an entire social group, he also maintained that religion developed in the same way and that it (always excepting the Christian religion for Vico) was the product of a social group. The early, basic mentality of these first groups took myths literally. Yet, at the same time, with mythical imagination there is always for Vico an implied act of belief.

In opposition to the common assumptions that the poets developed language and religion, Vico argued that new reasons must be established for the origins of poetry – and his *degnità* were to be the basis of this new approach to mythology and language. The presence of terms such as *voci mentale* and *dizionario mentale* was a clue concerning the shift in his discussion from actual or presumed language of the past and its development to the use of language as a means of historical reconstruction and to the linkage of ideas and language in his thought.

3. Language as an Instrument of Human Thought

All of Vico's interpretations of language depended on his view that it was not simply a means of communication but also an instrument of thought. Inasmuch as he argued that all languages evolved along the same lines, according to a rather inflexible pattern, it was environmental factors like climate which he used to account for the different stages of development in societies of roughly the same age.[1] Montesquieu was to write at much greater length in *L'Esprit des lois* (*The Spirit of the Laws*, 1748) about the importance of climate and geography in shaping individual societies and is generally credited with this theory,[2] Hume, in his essay 'Of National Characters' also published in 1748, trivialised the importance of climate and geography on cultural and national development. Its origins are to be found in Book VII of Aristotle's *Politics*.[3]

In his autobiography, after having nodded in the direction of Arab mathematics and philosophy (Averroes, 1126–98),[4] Vico still considered himself able to state that it was because he was born in Naples and not Morocco that he was a scholar. This most bizarre, not to mention inaccurate statement, though hardly unusual for its time, does not save Vico from relativism as much as it points out his own prejudices. In any case, it is of great importance in showing that Vico was not afraid to make judgements regarding past societies – far from it. He postulated that for the first time his *scienza nuova* allowed for critical analyses of the past. Developing social groups, with their codified myths, laws, were dealt with more or less impartially, but he did not attempt to avoid value judgements regarding more advanced societies. Vico recognised that the ability and confidence to make statements of fact regarding other societies led to judgements which inevitably involved some system of values.[5] Hence it is clear that, although Vico maintained that all societies followed the same basic pattern of growth and decline, he did not believe that all lifestyles, social arrangements and political systems were equal. He argued that his *arte critica* (critical art) established a basis for effective judgements of societies. In no way, therefore, can Vico be considered a relativist.[6]

The creation of early poetry was for him the first operation of the human mind. Thus the imaginative faculty of man was autonomous. Vico traced the development of primitive mentality from the knowledge of particulars to the grouping together by the imagination of these particulars. This then led to the creation of general types; next the other senses and faculties developed; and finally there was a division between imagination and reason.[7] Vico considered the human mind to be the same everywhere, having the same capacities, but not everywhere developed to the same levels. The issue was the degree of progress made.

However although early poetry was the focal point of Vico's attention, he nonetheless regarded poetry as an immature phase in the development of the human mind. (The scattered references to the development of the human mind are not discussed because of Vico's stress on early societies; the development of the mind is yet another intriguing topic sacrificed to his relentless pursuit of the theme of the evolution of cultures.) Vico gave primitive man no credit for disinterested thinking, beyond that of his primitive needs. The poetry of primitive man consisted in a vision of the world as a reality made and imagined similar to his own experience. Vico did not feel it was necessary to distinguish between early and theological language. Language, for him, was always poetic, never abstract, cynical or sarcastic. At the same time poetry provided the internal logic of Vico's work, the essential link among imagination, language and history. These elements together comprised his *mondo poetico* (poetic world).

According to Vico metaphors in early society were pervasive not only in language, but also in thought and in action.[8] The essence of metaphor was understanding and experiencing one kind of thing in terms of another. Far from believing that words have set definitions, he examined linguistic expressions as flexible containers for meaning. Metaphors were to him the conduit between experience and expression.[9] This view of the metaphor was in direct opposition to the seventeenth-century attitude which suspected the metaphor because it was connected with 'the false world of ancient superstition, dreams, myths, terrors with which the lurid, barbarous imaginations peopled the world, causing error and irrationalism and persecution' (paraphrase of M. H. Abrams by Berlin).[10] Yet as Berlin responded 'such ways of speech . . . only later became artificial or decorative because men have by then forgotten how they came into being and for which they were originally used'.[11] Vico was concerned with the primitive operation of metaphor in the evolution of speech. He declared that the poverty of words caused the origin of both metaphors and metonymy.[12] Vico's insight regarding metaphors, that they contained kernels of past experiences and emotions, has been noted numerous times. But the danger in such an emphasis is that the metaphor itself is seen as Vico's method, rather than simply one, albeit a highly useful, means to the past.

Ultimately metaphors – whether thunder or pagan gods – were insufficient to Vico. It was the stories, the myths, that he needed for analysis, and the metaphors were thus a part rather than the whole answer. The metaphor has very often been declared to be the key to Vico's thought. But it is more useful to see the myths, the stories themselves as the medium which transmitted the cultural information that he desired.

Vico assumed that language developed to deal with the problem of communication. He never discussed forms of communication between different societies, just as he never discussed culture-sharing of any sort.[13] It did not discomfort him that one word – *god*, for example – could have so many different meanings. Instead he wanted to analyse the combination of everyday assumptions and creative thinking which together fashioned living myths. For Vico language was never the 'mere clothing of a thought which otherwise possesses itself in full clarity', as Maurice Merleau-Ponty (1908–61) later wrote.[14]

Language was a human creation for Vico, and he thus believed the meaning of words to be a function, not a property, of terms. It was this aspect of language, that of conceiving and naming objects – which most concerned Vico; communication was secondary. Primitive forms of language were for Vico primitive forms of cognition. He discussed the concepts of ideas and language as almost interchangeable;[15] he stated that language was at the start of any ingenuity.[16] Language was, therefore, very much more than simply a means of communication for Vico; it was a form of thought. Further the intellect was that capacity of mind which enables it to conceive universal symbols and the meaning of ideas.

4. The History which Vico Sought

Following the tradition of Renaissance studies, the history which Vico sought to discover was closely bound up with philology and etymology. Philology was declared by Vico to be synonymous with both the history of speech and the history of human institutions. It was when he discussed his historical method of reconstructing isolated aspects of past cultures that he named philology as the particular means by which the origins and development of a society can be uncovered.[1] His new principles of mythology and etymology were to be used to examine the vocabulary of the first nations, which has been left in the form of fables.[2] Thus his *idea* of the new art of criticism was to be found in fables, and with his new approach

– which included etymology, philology and, most originally, mythology – the truth in pagan fables could be discerned.[3] Through much of his writings the terms language, poetry and philology were used interchangeably with the notion of history itself. Once again this practice demonstrated that Vico had a few central ideas which he discussed over and over in a myriad of ways.

Vico's account of the fundamental role of etymology, which he sometimes defined as philosophy and elsewhere as history, was the first systematic attack on Plato's dismissal of etymology as wholly useless.[4] In '*Nova scientia tentatur*' Vico delineated the need for the study of discourse and of variety in word usage.[5] Nonetheless he did not leave his readers any suggestions as to how such a study should be pursued, although in his defence it must be stressed that he was less concerned with the techniques of etymology or philology than with their results. Philology, Vico wrote in '*Nova scientia tentatur*', had to do with the history of speech and the history of human institutions. This statement is the first and one of the clearest of all of Vico's writings regarding the intimate relationship between language and history; the statement was also the basis of his concept of the 'common mind of the nations' ('*mente comune delle nazione*').[6]

In addition language was important to Vico in terms of his chronology. The study of language made possible the correction of the chronological table, which was the immediate purpose of the *scienza nuova*.[7] It is now difficult to appreciate the importance that Vico himself placed on his chronology, for he truly believed that it would demonstrate the validity of his new science. To what extent Vico meant this method to be applied to more recent historical artefacts is not completely clear. He maintained that articulate language, followed by philosophy, brought about the demise of imaginative symbolism. He abhorred the sophistry of contemporary intellectuals and would not have considered their deliberately hidden meanings to have been in any way as useful as early metaphors and myths, since such clever phrasing was the preserve of only a small segment of the population.[8] No doubt his desire for a larger reading audience was one of the reasons for his shift from Latin to Italian in mid-life, although he remained dependent on the educated classes to read, accept and teach his *scienza nuova* if it was ever to have any appreciable impact on society.

Vico cannot be considered an elitist on the basis of his theoretical works, his fascination with the aristocracy notwithstanding, for he was concerned with the the life and the way of thinking of entire social groups. On the contrary, Vico can be faulted on this point for not subdividing his societies even more, for his macro-approach with its all-inclusive scope came dangerously close to corrupting the unique, individual images of these past societies. In unsympathetic hands his method, even with the sympathetic interpretation and usage of his notion of imagination, could well blur the subtle and not-so-subtle social and economic distinctions which are present in every society.

Vico never gave any indication that he believed there was a correct inter-pretation of the past. The cycles, the chronology and all his new concepts (*scienza nuova, arte critica, chiave maestra* [master key]) notwithstanding, he never attempted (with the exception of the demonstration of the truth of sacred history) to demonstrate an optimum approach to the past or a single theme in history. His was a remarkably tolerant and flexible system – almost certainly more so than he intended. Far from being single-minded, Vico could well be accused of throwing his energies in too many diverse directions and, once again, of not leaving any specific advice on how to proceed with any of his theories in an orderly and coherent manner. He propounded the use of language and its complements, such as epigraphy, numismatics, and chronology (by which he meant his own scheme that listed by years all the great world civilisations), as well as philology, etymology and a great many other approaches to begin the task of comprehending past societies.[9] Even if it is ever true of other theoretical works, Vico's writings most definitely cannot be treated as a *vade-mecum*, a virtually infallible manual for historical studies. Just why Vico considered it necessary to comprehend history at all is not entirely clear. Historical awareness was not the essential element in saving societies from decline, according to Vico, nor would an awareness of history necessarily aid us in leading more fulfilled lives. However it is essential if we are ever to understand the creative abilities of any given time. What Vico did provide, however, was a new approach, a new way of thinking about history and historical studies. Research was no longer limited to topics that could be studied in official written records. Vico's work opened the way to alternatives to the political history which retains its dominance to this day. These alternatives clearly included cultural history, but by extension also intellectual, social and economic history.

Vico's work was revolutionary, not just in terms of content but also in its techniques. Anthropologists and sociologists have long since superseded Vico's suggestions regarding the means of exploring pre-literate societies.[10] But within the discipline of history, as well as for many other of the social sciences, it is the breadth of Vico's vision and his willingness to advocate almost any means – academic or otherwise – to achieve his ends, which are most inspiring.

The history that Vico sought to identify and that he discussed at great length throughout his writings was not the straightforward determination and study of past events, nor was it simply a system of pre-determined cycles. History to him was also much more than divisions or epochs; it was the testimony of those times.[11] The knowledge of the past which Vico desired was not to recreate specific events, primitive forms of government or hierarchy, but to recover the shared *mentalités* of the time, which all of the above examples could help to identify. Vico's historical method was a means of finding out about peoples in their social and physical environments, and in their particular place in time. His was a study of how people in earlier civilisations perceived themselves, their roles in society and their place in regard to both their ancestors and to following generations. Vico wrote that his approach was not just about *particular* early societies, it was to furnish all we will ever know about the birth of first societies. This was the *sapienza riposta* (hidden knowledge) that he sought.[12]

The ultimate goal for Vico of such a linguistic enterprise was not the production of a complete national history, or of a unified history of any given social group, much less that of a world history. Vico's history was a history of ideas. The rewards of such a historical reconstruction were neither concrete nor directly applicable but were much more personal. Vico set out the means of answering questions asked by many people at many different places and times. Vico maintained that it was possible to identify the particular categories of mythical thinking: these common notions, these wordless, unconscious attitudes dictated by social usage, that were at the heart of Vico's research.[13] His plan was to find the common ideals shared by all peoples, as set out in his *dizionario mentale*. Most importantly, such an analysis was the means by which historical knowledge of past cultures could be obtained and examined. Such a study was an investigation not only into past cultures but into the mind of the researcher. An investigation of history in this sense was a way of finding out about oneself and one's own culture; it was a way – and for Vico the only way – to comprehend the workings and development of the human mind.

5. Myths

La scienza nuova prima offered a radical shift in the study of language in European culture, for it was the first response to the issue of when and how figurative speech is born.[1] But Vico was to receive little credit at the time either for this insight or for his theory that Homer was not an actual historical person, but rather the culmination of the ancient Greek people, the concept which dominated the third book of the final edition of *La scienza nuova*. Vico did not intend his *logica poetica* (poetic logic) to be interpreted in a theoretical and abstract manner. On the contrary, one of Vico's greatest insights was that he recognised that poetry contained isolated images of previous cultures. Early poetry was for Vico one of the means to what we now refer to as cultural anthropology. For Vico metaphorical imagination (*l'immaginazione metaforica*) was the most useful instrument by which early man expressed his feelings and fears in analogical form, for example, 'the blood boils in my heart'.[2] His genius lay in his realisation that it was necessary to break the codes of these poetic characters and metaphors in order to make use of them in a cultural study of that time. These included, for example, the representations of virtue, strength and love by classical gods and goddesses and the description of anger and an awareness of the insignificance of man in comparison to nature by comparisons to everyday occurrences. Still, today it is often stated that only writing preserves the older form of a language, a generalisation which ignores the long centuries when language and culture were preserved in fables, myths and oral poetry.

Vico interpreted myths as a form of thought for a society. He considered it necessary to penetrate the mythical consciousness of a society in order to comprehend the way those social groups thought. The construction of the theory of myth was the *chiave maestra* to his entire system. He declared the mythmaker's mind to be the prototype, and the mind of the poet to be mythopoeic. In this manner he viewed myths as a form of intuition, both on the part of the societies and of the poets involved, as well as for the modern researcher trying to come to grips with these ancient fables. It was the genuine culture, the characteristic mould of a particular civilisation which concerned him. There was the emergence of a concept of personality of cultures – for example, ancient Rome as war-like – in his writings. In the words of Susanne Langer (1895–1985) 'every society meets a new idea with its own concepts, its own tacit, fundamental way of seeing things; that is to say *with its own questions*, its peculiar curiosity.'[3] Although Vico was concerned with the dynamics of myth preservation and change, it was actually the nature, content and form of the earliest versions of these myths that he sought.[4] Yet it must be noted that Vico's work was limited somewhat by his belief in pure language. He had relatively little interest in the development of language; rather he wanted to strip away all the intervening layers.[5]

Myths tend not to be logical in their arrangement, nor internally consistent. This issue was not important to Vico, for it was the content of the myths which was the essential element to him, the attempt to deal with the origin, shape, function and destiny of the world. The same themes recurred in many parts of the world. The relationships between gods and men, men and spirits, men and men were discussed in rich detail. Many myths described a primitive, glorious and happy state in which men and gods once lived together in perfect harmony until something tragic happened. The myths were an attempt to deal with personal problems such as illness, as well as cosmic events and natural phenomena such as earthquakes, lightning, thunder, rain, frost, eclipses and other catastrophes.[6] Vico was not at all concerned with the idea that myth does not give man any power over nature, only with the illusion that he understands it.[7]

Myths served an explanatory function in society, answering fundamental questions about death and the meaning of life. These central topics recurred over and over in primitive myths and in early poetry from every part of the world. This tenet of Vichian thought is constantly reaffirmed by the research of modern social anthropologists. Echoes of Vico's views can be also be found in the writing of modern linguistics. For example: the primitive mind was coarser than our own and completely determined by basic needs (Bronislaw Malinowski, 1884–1942) and differences between modern and primitive thought were inevitable because the latter was entirely determined by emotion and mystic conception (Lucien Lévy-Bruhl, 1857–1939).[8]

The categories which Vico devised for the analysis of myths have not lost their relevance even in our age of specialisation. For example, one can find in Vico all four of Joseph Campbell's (1879–1944) later functions of traditional mythologies.[9] The first was the reconciliation of the consciousness of a society with the preconditions of its own existence. Campbell wrote that a society attempted through myths to redeem human consciousness from its sense of guilt in life; Vico wrote of shame (*pudore*) as the motivating factor which civilised primitive man and led him to express these feelings in fables. This mystical function of myth was followed by a cosmological function: myths were all important to a primitive society in formulating and rendering an image of the universe. Vico discussed at length the explanations of early peoples for their physical surroundings. Just as important was the sociological function, which was responsible for the validation and maintenance of the individual social order. In this way, Vico attributed the taming of primitive man to early poetry.[10] The final function was the psychological – shaping individuals to the aims and ideals of their various social groups, bearing them through the course of human life. Thus to Vico myths were responsible for the social cohesion of early societies. They gave a group of heretofore disparate peoples a way of thinking about their roles in regard to the supernatural, in nature, in society and as individuals.[11]

These stories had a fundamental function in society itself. A myth was a conceptual statement about man, his society and his universe and generally was the story of acts involving the supernatural. The myth itself was considered by that society to be true and sacred, and was related to a creation, how something came into existence, or how a pattern of behaviour, an institution or a manner of working was established. Myths dealt with the origin, behaviour and destiny of man; they were an attempt to account for the origin of the world (and the local tribe) even in pre-literate societies. Thus Vico viewed myths as paradigms for all significant human acts. The mythopoeic need for myths in these early societies was based on the belief that by knowing the origin of things, one could control them at will. It was believed that by recounting the myth it lived again and the events which it recounted could be re-enacted.[12]

6. Social Institutions

It could be argued that language was not Vico's best developed theme and that he tried to discuss too many diverse topics under this heading – the development of language, early folk tales, myths and codified laws. A cursory reading of Vico could lead one to believe that he was more interested in language than in society. But time after time Vico himself leads us back to his main theme of history, and there the complementary role of language is made clear. The phrases *sapienza volgare* and *logica poetica* were used almost interchangeably. Vico glorified the wisdom of early man not because he considered that he had reached any higher intellectual heights – perhaps because he had greater awareness of his physical and social environment – but primarily because early fables and myths, now preserved in a much more polished form, were all that was left of a forgotten time. A prolonged reading of Vico leaves one with the feeling that there must be another reason as well: that something is lost at every stage of development and that to gain a more complete comprehension of the ebb and flow of historical development, if not also for a more personal reason as pursued by the Romantics, it is necessary to try to retrieve these lost attitudes and mentalities. Something happened at each stage of development of a society without which the most valued contributions of that society could not have been made.[1] For example, Vico claimed that only the cruel society of the ancient Greeks could have produced the *Iliad*. Thus he teased his readers by reminding them that (in his opinion) the most brilliant literary and philosophical works were produced in societies which one cannot always reproduce or approve.

It was the principle of civility (*ratio civilis*) which offered the clue to unravelling the myths. Vico stated that the study of language is called *humanitas* ([Latin], *umanità* [Italian] – culture, humanity), because it is affection that induces men to help each other – particularly those who speak the same language.[2] Humanity, and hence human nature itself, were thus inextricably bound by language. This is, according to Vico, another reason for the study of language – that one is thus analysing particular social groupings formed along linguistic lines. The issue that language groups were often synonymous with nations, or at least that the desire for this to be so was present, was not lost on Vico. For him language was inseparable from humanity: it was language and not a mind capable of rational thought that makes us human rather than bestial. Yet his view could well be countered by the argument that language is just one of the great many functions of which the human mind is capable; that it is a part rather than the whole. In this way he used the term *lingua* to describe all human and humane qualities.

Vico wrote that *humanitas* was composed of shame (*pudor*) and liberty (freedom, a way of thinking befitting a freeman, *libertas*), which together were the roots of liberality (generosity, a way of thinking befitting a freeman, *libertas* – (*Il diritto universale, De constantia iurisprudentis*, Book II). This was one of the most sophisticated mixtures of Vico's concepts. Shame, the negative force which civilised primitive man, and liberty, a concept not discussed in a theoretical manner elsewhere in his work, together helped (we are not told the other elements) to comprise liberality, which he used to denote the attitude of those who freely aided others similar to themselves. Such an attitude of cooperation was mandatory for the transition of family units into villages, cities and then to nation-states. Vico considered growth to be natural and, thus, gradual. He also discussed language in society as an educational tool, an active force in the taming of primitive man, because poetry, in his view, taught the vulgar to live piously. Hence he defined a social role for language in which the latter was intimately involved in the taming of primitive man. Myths and stories passed orally from generation to generation, instilled in listeners a sense of religion and of respect for superiors, and fostered the development of creative imagination as manifested by the capacity for constructive and mythological thinking. All these factors were essential to the process of turning wandering peoples into monogamous cultivators of the land. In his own time Vico accepted that not only mathematics, but – more significantly in the scheme of his own work – rhetoric formed a basis for education. Rhetoric combined many of the skills he considered most pertinent for the civil education of the young: memory, linguistic ability and assertiveness. Thus language had an educational function not only in the early stages (via the elegance of early poetry), but, unlike many of Vico's other favourite topics, throughout the development of a society. For Vico the benefits of a language, by which he meant its civilising force, were shared by all people. Yet there was nothing of the missionary in Vico. His desire was not to spread his interpretation of language to primitive groups of his time, such as the North American Indians, Chinese and Japanese, whom he discussed at various points. Quite to the contrary, he saw language as a means of analysing these societies as they·

were; he had no desire to disrupt the balance of their social development. Nor did he foresee practical benefits deriving from his work, no philological or etymological theories which would help in the study of specific ancient or otherwise unknown languages.

Instead Vico desired to find the order in myths for the purpose of determining the social structure of early cultures. This view was expressed in this century by Claude Lévi-Strauss: *'Qui dit homme, dit langage, et qui dit langage, dit société'* ('He who says man, says language, and he who says language, says society').[3] Myths were responsible for integrating the universe, natural and supernatural, into a form understandable by each fully-functioning member of the community.[4] Vico wrote:

. . . che sulle cose le quali si meditano vi convengono le nostre mitologie, non isforzate e contorte, ma diritte, facili e naturali, che si vedranno essere istorie civili de' primi popoli, i quali si truovano dappertutto essere stati naturalmente poeti.	Our mythologies agree with the institutions under consideration, not by force and distortion, but directly, easily, and naturally. They will be seen to be civil histories of the first peoples, who were everywhere naturally poets.[5]

Myths were involved in renewal on many different levels. They re-enacted creative events of the past; many of the stories had to do with the renewal of health or even life. Perhaps most importantly, myths helped men transcend their limitations by lifting their spirits and explaining their place in the universe. Myths were the means of evaluating, classifying and relating all of life's experiences. Thus the making of myths, this universal human trait, was essential for the preservation and transmission of the beliefs and attitudes of individual societies.[6]

Myths validated social behaviour -- rules of behaviour, mores, taboos and attitudes were all confirmed by mythology. Mythic injunctions served as a basis for parental or community demands. In this way myths validated beliefs, and beliefs were in turn reinforced by everyday events. Myths also served as a model for ritual. Conscious repetition was necessary to keep the details of the stories fixed in the mind. Changes were made in the stories, sometimes based on group consensus, or by reinterpreting old elements of the stories in terms of new information. Eventually these modifications led to the restructuring of the myths themselves.[7]

Myths were used by Vico as a way to go beyond texts and apprehend the history of cultures directly. There is no doubt about the fundamental importance of culture to Vico, for he viewed thought and culture as synonymous. He was also most aware of the interdependence in society between culture and the framework of the state.[8] Nonetheless Vico was not attempting to develop a history of comparative cultures but of human social institutions. Indeed he cared less about what was created than that it was created. For Vico history was man's creation, and he desired to identify as many aspects of human creativity exhibited in developing societies as possible. However in his rush to define and delineate the concept of culture and individual societies, he did not fall into the trap of ethnocentrism. *La boria della nazioni* (the conceit of nations) was strongly condemned by Vico. This sort of easy generalisation based on cultural conditioning was for him neither justifiable nor useful.

7. Order in Myths

Three main approaches to myth analysis have developed since Vico's time. The historical school regarded myths as half-forgotten historical events, imaginatively embellished tales in which deities were merely magnified men.[1] In this manner myths represented at the very least non-historical reality. Although this was not Vico's approach, there is a parallel here with his view of human language as a copy of divine language.[2] And one can find in Vico strong echoes of the classical theory of myth as an allegory of philosophical truths which inspired Bacon's *De sapientia veterum*.[3] Yet he was if anything in closer accord with the modern psychoanalytical approach in which myths are viewed as externalised wishful thinking. The material of these early myths was similar to the symbolism of dreams – image and fantasy. According to Vico myths were public dreams, representing the hopes and fears of a community, but he never pursued this line of thought to conclude in the manner attributed to Sigmund Freud (1856–1939) that dreams were private myths.[4] Still Vico would have been in accord with Freud on the point that childhood, the period of our lives which affects us most as adults, is paradoxically the one we can remember least well. Vico's aim was, clearly, to reconstruct these half-forgotten first years and stages of a society.

In the third, psychoanalytical, approach, myths are viewed as the foundation for understanding interpersonal relationships and most importantly as the means to gain understanding of individual and group behaviour. This type of access was considered possible because language had developed to solve practical problems. More important than the individual motifs (symbols or metaphors, for example), which were very often the same in any society at the same stage, was their overall arrangement. This structure of beliefs gave crucial information regarding the world view of these early societies.[5]

Vico maintained that there was an inherent order in mythological stories, and he denied that he was forcing his own theoretical structure onto them. He saw no essential difference between mythological thinking and that involved in the writing of history, in that they both attempted to portray the spirit of a given culture at a given time. Thus it appeared entirely natural to him that the telling of a myth was done in innumerable and diverse ways, which very often transformed the original story. He viewed epic stories as broken fragments of an on-going saga. He did not consider the disorganized state of the myth to be the initial form.[6]

Vico asserted that there was no exact point where mythology ended and history started. The gaps between myths and history should be bridged by histories which are conceived not as separate from, but as continuations of, mythology. Yet myths, as histories, were extremely repetitive. Thus it was by the sifting of myths that one comes to a better understanding of historical sciences.[7] By this point history had replaced mythology, since it fulfilled the same functions. For example, historical accounts in the eighteenth century were almost entirely based on written documents.[8]

Vico recognised that these cultures could never be regained exactly as if they had not been lost. Yet he stressed that there must be an awareness of their existence and importance. Vico wanted to find the order behind the disorder in history, by identifying the common rituals and patterns in the development of societies. He stressed the similarities rather than the differences in the development of cultures. He saw that a study of myths of very different cultures can close some of these divisions, explaining apparent discrepancies.[9]

His discussion of language and mythology was an attack on the objective approach to thinking and society. Vico strongly disagreed with the long-standing belief that words have fixed meanings, and departed from the supposedly enlightened attitude that language should be as direct as possible, and that poetic, fanciful, rhetorical and figurative language should be avoided. At the same time he asserted that one experiences the world by acquaintance with objects in it, one understands these objects in terms of concepts and categories, there is a reality which is not subjective, and one can say things that are absolutely, impartially and unconditionally true and false about it. This final point was strengthened by the stress he put on *senso comune*, the shared values of any and all social groups. He believed that *senso comune* could be illustrated in either a slightly or very distinct manner in each society, but that because of these shared values that all societies at all times have in common there was some basis for comparison.[10]

Vico's approach to life and language was not completely subjective. Although he recognised that in most of our everyday activities we rely on our senses and develop intuitions we can trust, Vico did not consider the most important factor in one's life to be one's feelings, or that poetry transcended rationality.[11] Nevertheless he declared that the language of imagination was necessary in order to express the unique aspects of our experience.

Vico's originality lay in his view of language as a way of thinking, as both the expression itself and as part of the actual, creative process. The critical importance of language to Vico was not only that it was a human creation, but that it was the means of discovering other human creations. Language was the identifying mark of a people, and, as with customs, it would develop in its own particular way in each society.

Vico had five main aims in his examination of language and culture, of speech and society. First, he desired to find out the mentality of early cultures. This led naturally to the second point, which involved an analysis of developing social structure. Third, he recognised religion as part of the social fabric, permeating all aspects of early society, as can be demonstrated from myths. Fourth, Vico maintained it was his *dizionario mentale* which allowed exploration of the past. Fifth, it was *fantasia* which worked through language and myths, and which made possible the expression of individual cultural characteristics and interests. The crucial relationship between imagination and language has long been neglected, in part because language was viewed as the only creative force. Vico's plea for the recognition of imagination in human society was echoed in this century by Michel Foucault (1926–84) who argued that the insane and irrational elements have always been excluded by European culture.[12]

8. Imagination and Historical Reconstruction

Vico's overwhelming emphasis on language rather than rites of religion, as the best means to study societies of the past, is vindicated when his dependence on language as the proper tool for the application of his critical art is recognised. It was by means of the discovery of the true nature of poetry that he demonstrated the key to *La scienza nuova*, that is, imagination.[1] According to Vico, myths were faithful records of true narration (*vero narratio*), since they had to do with both the imagination and the will of the community.[2] He would not have been distressed by records which were written deliberately to deceive, for he would have considered them to have been true to their basic purposes. Unfortunately he gave no clues as to how we can detect deliberate falsifications in historical documents.

There is an essential point here: Vico himself sensibly realised that it was necessary to grasp not only the imagination but also the will of a society before a true understanding of that culture could be gained. Hence his was an attempt to recreate not only the transient feelings of a people, but also an effort to understand what types of people they were. Ultimately Vico's distinction between the will and the spirit of a people differed very little, no doubt accounting for why the concept of the will does not appear in his later writings.[3] But the differentiation of the two in *Il diritto universale, De constantia iurisprudentis* gives us an important insight into his definition of imagination, helping us to understand better his own personal vocabulary.

Collingwood put forth the interpretation that Vico literally meant that we should use our own imaginations *entrare* and *descendere* into the conditions of those people in the past we should like to know.[4] There is a parallel here between Freud's belief regarding reason and Vico's imaginative historical reconstruction. According to Freud reason cannot save us, nothing can, but reason can mitigate the cruelty of living. A similar analogy would apply to Vico and the historical knowledge acquired by means of imagination.[5] However unfashionable this Vichian view, introduced to the English-speaking world by Collingwood, might be today, many history teachers at all levels still urge their students to do just this. But using the knowledge we have gained through a linguistic study of a society, based on the codified versions of their myths and laws, how does one begin to incarnate the skeleton of a lost civilisation? Vico, and to a greater extent Collingwood, was at fault for not sufficiently stressing how difficult and limited such an historical reconstruction based only on oral records would be. For the results might be utterly inaccurate and there is no way to verify them. But the difficulties in such an approach, the impossibility of shedding the conscious and unconscious prejudices of our own culture, should not be sufficient to deter us from trying, particularly as this is very often the only means possible to recapture lost times. This is the third usage of imagination in Vico: we avail ourselves of our own creative powers and intellects in this attempt at historical reconstruction.

As has been demonstrated, language in Vico was much more than a means of communication or an instrument of thought; it also actively shaped social development from the earliest to the most advanced stages. Language developed according to particular physical environments, but it itself (in the form of myths, legends and poetry) shaped and was shaped by the social world and the visions of life and nature of particular groups. Language for Vico had profound implications in terms of historical reconstruction and techniques.

Although Vico called poetry a gift from God, born of curiosity, he discussed it as an innate instinct, fashioned by environment and ultimately a human creation.[6] These seemingly contradictory characteristics demonstrate the crossroads at which Vico was standing. His statement that poetry was a gift from God added nothing original to his work, although this is not to say that Vico did not believe in both concepts himself. The birth of poetry by means of curiosity confirmed the place of imagination, while at the same time the nomination of ignorance as the mother of poetry reinforced the fundamental role of poetry in the earliest phases of a culture.[7] Both of these metaphoric relationships stressed the origins of language in response to environmental conditions. Language as an innate instinct in Vico had more to do with *il dizionario delle voci mentale* than with poetry as a divine gift. Throughout Vico's writings it is his preoccupation with a rich mythological universe, and most particularly with language and society as human creations, which is most evident and inventive.

5 Imagination and Historical Knowledge

1. Introduction

One can hardly fail to notice in Vico's writings the quite amazing jumble of ideas and terms. The danger when one is presented with such an inchoate mixture of the startlingly new and the prosaic is to read into it an interpretation which is not dissimilar to that held before reading Vico. But even if Vico's lack of precise terminology is to be greatly regretted, the continuation of such an approach would be intolerable. There is no doubt this has hindered a proper understanding of the Vichian view of *fantasia*, for much of the confusion regarding Vico's views on this pivotal issue is because it has been relegated to a subordinate position, as simply the means of creating mythology and early poetry. Thus the only hope of understanding the profound implications of Vico's discussion of *fantasia* is to begin by dividing into categories the three main ways in which he used this not uncommon term.

The first is *fantasia* as the attempt by primitive peoples to make sense of their physical and social environments.[1] Virtually every writer on Vico in the last thirty years has mentioned this important point, but without exploring how it is related to Vico's views on historical knowledge in more than the most general terms. With only two or three exceptions, this has been the only aspect of Vico's discussion of *fantasia* which has been recognized by scholars in the field.[2] However Vico used *fantasia* in two additional senses, the second being the spirit of a particular age.[3] It was not only the component parts, these attempts to explain human existence through poetry and myth, but more excitingly, it was also the composite mentality of the people of a particular civilisation at a particular time which was Vico's major concern. Vico had no desire (nor did his method attempt to provide a means) to reconstruct past events, much less to describe an ideal state or to provide a political system which he desired to see implemented in the years to come. Rather he sought to gain historical knowledge by comprehending the ways of thinking and feeling in these early civilisations. The third usage of the term describes the function we must ourselves employ to unlock the minds, the consciousnesses, of these past civilisations to reconstruct patterns of past cultures – this is truly Vico's *chiave maestra*.[4] Little attention has been paid to this critical, final point except by Verene and those, like Dray and van der Dussen, writing on Collingwood.[5]

Vico's approach to imagination was characteristically broad in scope. Nowhere in his writings do we find a close examination of the exact relationship between imagination and the development of the primitive human mind. Yet in virtually all of his theoretical works, and as early as the orations of 1699–1707, this connection is among his most dominant themes, one which he usually demonstrated by means of his discussion of the development of a civilisation. But Vico's intention was never to examine the contents of the human mind in the manner of Locke; instead he was concerned with the manifestations of this mental development as shown by the gradual growth of a society. It was imagination as the creative human and social faculty which occupied his complete attention.

2. *Fantasia* as the Poetic, Recreative Instinct

The traditional interpretation of Vico and his views on *fantasia* was reflected by Giulio Lepschy in a fourteen-page article in *Lettere italiane* (1987) entitled '*Fantasia e immaginazione*', which devoted only seven lines to Vico, and referred to him simply as the precursor of Romanticism, Idealism and Croce.[1] Sadly, among Italian scholars both outside and inside Italy, as in the English-speaking world, the emphasis is still on relating Vico to wider European trends. Thus Vico's more original insights regarding imagination have been ignored for the last two and a half centuries, primarily because they have no obvious parallel with those of a better known thinker.

Yet as early as his orations of 1699–1707 and again in the following year when he presented *De nostri temporis studiorum ratione* as the grandest and most ambitious of his inaugural addresses at the University of Naples, Vico spoke of the fundamental importance of imagination.[2] In *De antiquissima italorum sapientia* he discussed the division of imagination into two component parts: *ingenium* (the power of connecting separate and diverse elements) and *phantasia*.[3] Recognition of this relationship is crucial for a proper understanding of imagination in Vico's philosophical works; because in this way *ingenium* and *phantasia* were not distinct abilities competing with each other, but instead were constituents of imagination. Vico averred that *ingenium* was to be applied in practical spheres such as mathematics, astronomy and mechanics. He used the terms in such a similar way that it would be pointless to speculate whether *ingenium* could be applied to the arts and *phantasia* to the natural and physical sciences.

In *De nostri temporis studiorum ratione* he stressed the importance of imagination and memory in a pedagogical sense,[4] while in *De antiquissima italorum sapientia* he listed several examples of the fruits of *ingenium*: mathematics, design and the most colourful one – the pyramids.[5] He wrote in this oration that imagination is a true faculty because of the creation of images (*'Phantasia certissima facultas est, quia dum ea utimur rerum imagines fingimus'*),[6] leading to his even better known discussion of *verum* and *factum*. The second half of *Il diritto universale*, which begins with *'Nova scientia tentatur'*, finished in 1721, stressed over and over Vico's fascination with what he termed as the 'magnificence of imaginings' (*'Imaginum granditas'*).[7] In the same work he argued that imagination was the result of man's poetic faculty, which disappeared in the later phases of a society as the sciences developed in strength, though he was not at all specific about when or even at which stage it disappeared. According to Vico imagination, in the first sense of the poetic instinct of man, withers away completely in a society which has gone on to the study of philosophy and mathematics. Once again, the only concrete – but very eccentric – exception he ever gave, in which some remnant of the primitive form of imagination would remain in a later stage of a civilisation, was the consuls in Rome and their role as preservers of the law and thus of the earliest version extant of their language.[8]

Vico did not always clearly distinguish between his discussions of imagination and those of memory. In *De antiquissima italorum sapientia* he wrote, 'Men can remember nothing not given in nature' (*'Homini fingere nihil praeter naturam datur'*).[9] The implications of this quotation are profound, for consciously or not Vico has denied any supernatural element in the development and functioning of human thought. Memory is thereby separated from *fantasia*, which allows for invention and fiction. In 1744 he wrote that 'Resemblance is the mother of all discovery' (*'Similitudo mater omnis inventionis'*) and that men judge the unknown by the known (*'ch'ove gli uomini delle cose lontane e non conosciute non possono fare niuna idea, le stimano dalle cose loro conosciute e presenti'*).[10]

As discussed in Chapter 2 in the sections on *De nostri temporis studiorum ratione* and *La scienza nuova prima*, Vico generally did not make value judgements regarding imitation, which he discussed in terms of the creative arts (*De nostri temporis studiorum ratione*), laws (*Il diritto universale)* and fables (*La scienza nuova prima*).[11] He viewed the process of recognition and differentiation as central to the development of the human mind and of societies. In the 1730 edition of *La scienza nuova* Vico discussed the critical role of the rational faculties as a complement to his favourite topics of imagination and memory:

Ma tutte queste, anziché pruove le	But all these, rather than

| quali soddisfacciano i nostri intelletti, sono ammende che si fanno agli errori delle nostre memorie ed alle sconcezze delle nostre fantasi, . . . | proving to the satisfaction of our intellects, amend the errors made by our memories and the disorderliness of our imaginations . . . [12] |

For all the stress Vico laid on imagination he did not trust it entirely as the sole means of historical construction. His reluctance to rely wholly on imagination is in line with his conviction that the facts one obtained from myths might be completely inaccurate, but the attitudes could never be falsified – for example a direct contradiction of the facts (if indeed this could be proved) would itself reveal a crucial emphasis on a particular issue.

Far too often *fantasia* is simply equated with mythology in Vico's work, but *fantasia* is best viewed as a faculty or an ability rather than an art or a learned method, much less as a product of such an art of which myths would be the prime example. Vico argued in *Il diritto universale* that the poets were simply conduits of the knowledge of a society, usually that of the generations preceding them.[13] Thus it was certainly not the wisdom or the *fantasia* of the poets but of the people which was passed on. It is essential to note that Vico never appears to have used the term *fantasia* in a negative sense. Even if inaccurate, he always saw it as constructive and absolutely fundamental to the progress of a society. He never viewed it as a characteristic of stagnation, due to its repetitious quality: parent to child and poet to the people. In *De nostri temporis studiorum ratione* Vico advised teachers in no uncertain terms to develop memory in their students in a systematic manner, because it was almost *identical* with imagination.[14] From the first orations, Vico argued that imagination had not died out in his time, and that the modern variant of imagination was memory. Or, more exactly, he assserted that imagination had completely diminished in his age of science and philosophy and that memory was the closest one could come to imagination in this later stage of his own society. Either way, the preservation and cultivation of such knowledge by means of memory was of the greatest importance to him.[15]

Vico wrote that it was curiosity which led early man to explore his environment,[16] and this issue was central to his ideas about language, the development of the human mind and history itself.[17] For Vico language was natural, the spontaneous expression of human thoughts and feelings.[18] Language was not invented by the philosophers.[19] Although he maintained that rational thought was the apex toward which all mental development was aimed, he was nevertheless emphatic that the more sophisticated modes of human cognition were not necessarily superior in all ways to the primitive attempts at understanding their physical and social worlds. Vico held that only when the desire to push back established boundaries of knowledge and learning was present could further additional knowledge or insight be gained. He did not contend that this desire had to be fully articulated or even conscious, but simply that the demand for additional information was in itself the vehicle by which the human mind progressed.

Vico called the fables of primitive peoples *sapienza* (wisdom).[20] Yet he certainly recognised that these people did not have the educational background or technical expertise to explain natural phenomena even to the level of sophistication practised in his own time. This was of no concern to him, since he viewed myths not as an inferior version of modern wisdom, but as constructive attempts to resolve problems of human existence and social organization. Unlike Croce, Vico himself had no real interest in studying or collecting local fables and folk songs.[21] Vico's great interest in mythology was theoretical and functional in that it was provided a means, and he was persuaded the only means, to discover the ways of thinking and feeling of past civilizations.

3. *Fantasia* as the Expression of the Spirit of a Particular Age

The second major use Vico made of the notion of *fantasia* was as the description of the spirit of a particular age.[1] It was the same faculty which produced poetry (or religion or any other social institution) and the culture of a society. Berlin was exactly right that it is at this point we see 'the emergence of the concept of the uniqueness and individuality of an age, an outlook, a civilization'.[2] This aspect of Vico's work is generally ignored altogether, unless it is linked to a discussion of Vico as a precursor of the Romantics or Hegel. Writers on Italian literature are often the only ones to recognise the importance Vico put on this aspect of *fantasia*. An isolated example of this was J. G. Robertson's chapter on Vico in his 1923 book entitled *Studies in the Genesis of Romantic Theory in the Eighteenth Century*, which is better than anything written on this connection since.[3] This second role which Vico assigned to *fantasia* is not at all subsidiary – on the contrary, this was exactly the type of historical knowledge which Vico craved.

Vico's importance is primarily that he recognised the value of the early stages of any society and that he had faith that it was possible to tap the early wisdom of that time even in the later stages of different societies. Vico suggested that the human mind was active long before it reached rational thought, although he never denied that the earlier was a lower stage of conception.[4] Without glamourizing these early men or their achievements, Vico recognised that something is lost at each stage as a society develops and that only by comprehending these very early phases can one gain proper historical knowledge of a particular age's way of thinking. Vico's writings could perhaps be best described as a philosophy of the history of culture. According to Vico imagination in the first and second senses discussed above could only be manifested by a culture, the aggregate of individual attempts at self-expression. There is no indication that he considered there to be any limit to the number of forms by which imagination could be expressed. He clearly expected it to vary dramatically culture by culture and age to age and that these disparities comprised the identifying marks of a particular period.

Vico showed no great interest in planned social action, yet its creative counterpart, collective imagination, was the focal point of all his theoretical works. Social change most certainly occurred within the Vichian framework, but gradually, not as the result of single actions. Perhaps for this reason there is no discussion of social and political responsibility except in terms of education. For Vico the outcome of a civilisation primarily had to do with decisions made by that social group. Again this supports the view that there is no evidence that cataclysmic decline was programmed into his system, but rather that he recognised its repeated presence. Nor is there any evidence that Vico considered his theory of history to be either mechanistic or static. His historical cycles undoubtedly affirm that he accepted any given society to be part of a continuum. Yet in no way would he have believed that this excluded the possibility of novelty.[5] This was seen in '*La pratica*', in which he intimated that proper moral and civic education of the young could avert the third stage, the final downfall of society – thus destroying the interpretation that Vico accepted the character and health of cultures to be predetermined. Indeed the whole tone of his theoretical works stressed the vitality, spontaneity and originality of the early stages of each society. It was, without a doubt, by means of a study of imagination, and thus of the creative instincts as manifested within particular societies, that Vico asserted that one could obtain the proper historical consciousness.

4. *Fantasia* as *la scienza nuova*

Vico's concept of historical consciousness encompassed a group's growing awareness of itself as a separate, legitimate social unit and a parallel transition of mental processes from imagination to rational thought.[1] Vico maintained that the development of the human mind was progressive, even though this clearly clashed with the Christian view, and he also contended that the essential nature of man was itself changeable.[2] He asserted that the history of past civilisations could only be interpreted if one could understand the human nature of that time. In an era when Neapolitan intellectuals tended to align themselves with the scientific approach, Vico constantly reaffirmed the complexity of human nature. In the words of Elio Gianturco, for Vico human nature was composed not only of 'sheer rationality, nor merely of intellect, but also of fantasy, passion and emotion and his insistence on the historical and social dimensions'.[3] Indeed Vico's emphases were always on the historical and social dimensions. There can be no doubt that Vico would have argued that the development of the human mind through imagination could not have been separated from the development of human society. He regarded historical consciousness as developmental, and the potential for greater awareness of the past as increasing with the progress of a society. He put forward a case for a higher level of historical consciousness than any of his contemporaries and maintained that his readers could benefit directly from his insights by making use of his method.[4]

Vico held that analysis of the development of human consciousness (*umana mente*) was the single most important part of his work.[5] Perhaps the second best known aspect of his writings (after *corsi e ricorsi*) was that he argued it was possible to comprehend history because it was made by men. This statement can only be true if there is some means of at least penetrating the primitive psyche of man and if the primitive consciousness held the same basic notions of *humanitas* that were considered innate in his time. Without this latter point in common, one might as well study a different species.[6] Clearly it is here that his *dizionario di voci mentale* delineated in *La scienza nuova prima* played its most crucial role.[7]

In the final version of *La scienza nuova*, *fantasia* is used in two distinct senses: there is the *fantasia* which is the means of the creation of poetic wisdom, and there is the *fantasia* which functions as the medium through which understanding of past societies can be gained. Vico was not unaware of the very different uses he made of the concept of imagination. At the end of Book I of the 1744 edition in the section on method he wrote:

Finalmente, quanto gran principio dell'umanitá sieno le seppolture, s'immagini uno stato ferino nel quale restino inseppolti i cadaveri umani sopra la terra ad esser ésca de'corvi a cani . . .	Finally (to realise) what a great principle of humanity burial is, imagine a feral state in which human bodies remain unburied on the surface of the earth as food for crows and dogs.[8]

In this example Vico challenged his reader to make use of his own faculty of imagination in order to comprehend this early society, which for once was not described in heroic terms as being strong in imagination, but rather was presented as an example of the anarchic quality of mankind without the civilising force of a coherent society. It should never be forgotten that the complete title of this work is *Principj di Scienza Nuova di Giambattista Vico d'intorno alla comune natura delle nazioni* (*Principles of the New Science of Giambattista Vico according to the Common Nature of the Nations*).[9] His new method of *fantasia* was not simply to be applied to individual aspects of human nature, such as shame. More correctly, it was the dynamic aspect, the tension, among the diverse components of a society which Vico viewed as new and interesting.[10]

Vico's aggressive attacks on the conceits of both scholars and nations are more comprehensible in the midst of his study of ancient societies if viewed as a threat, in his eyes, to the proper use of imagination.[11] As discussed above, as early as 1707 Vico wrote that the young should spend some of their school years uncontaminated by the sophistry of established curricula so as to preserve their child-like imagination.[12] This is one of Vico's most explicit attacks on established pedagogical and historical practices of his day. *Boria*, which implies arrogance and complacency as well as the usual translation of conceit, was confronted by Vico in a place of prominence (Book I, 125–128) in the 1744 work.[13] National conceit, *la boria delle nazioni*, was contemptuously dismissed by Vico, for it undermined the whole purpose and proper function of historical studies.[14] Scholarly conceit, *la boria de'dotti*, the assumption that all the people in the past were themselves scholars, with the background and orientation which that entails, also received scathing treatment from Vico for its one-dimensional approach to other cultures and times.[15]

Imagination was for Vico the basis of human historical reality. Far from devaluing the importance of *verum* and *factum*, imagination complements this much better known aspect of his work. There is no contradiction in his thought between the two concepts; indeed his belief that one could only truly know what one had made, '«*verum*» et «*factum*» *reciprocantur . . . convertuntur*', was fulfilled by means of imagination, the creative instincts of man. This is one of the strongest links regarding the intimate relationship between imagination and historical knowledge in Vico's thought.

5. History

One of the reasons that Vico used early societies as his model was because he maintained that they were more likely to recur in a relatively similar form in various cultures.[1] He considered them to be the basic form of society, uncontaminated by education and training. This is not to suggest that Vico viewed the first, primitive cultures as unchanging. Vico was fascinated by the permanence of *change* in human societies. He recognized the extremely close relationship between local and national history and it was the unconscious elements in society which he sought. Vico insisted that the interplay between the conscious and the unconscious elements of history provided the means to discover the 'course the nations run'.[2] The role of unconscious elements in history, particularly in psychohistory, is still contentious today. But Vico contested the opposite viewpoint. At the time he wrote, only divine providence was granted a role in history which was impossible to identify in any precise fashion, albeit its presence was considered to be both obvious and fundamental.[3] But Vico considered it necessary to recognise not only divine providence, but also to identify the ways of thinking and feeling of communities, since he believed only then would it be possible to understand the events and patterns of the past.[4] According to Vico in *De antiquissima italorum sapientia*,

Historici utiles, no qui facta crassiuset genericas caussas narrant, sed qui ultimas factorum circumstantias persequuntur, et caussarum peculiares reserant.	. . . useful historians are not those who offer imprecise accounts of the facts and generic causes, but those who search for the ultimate circumstances of facts and and search for particular causes.[5]

Vico was engrossed in what is now termed the cultural unconscious and in its transmission.[6] It was the development of belief systems and other social forces which Vico most wanted to trace. In particular he was extremely concerned with culture-bound elements, for these were the very factors which distinguished a particular age or each successive culture.

Vico considered language to be intimately related to the general conception of a cultural system. He asserted that in its first stages language was essentially amoral (in terms of relations between individuals) for it reflected the sentiments and prejudices of a single social group. Yet as that same society developed, its language moulded later generations in their ways of thinking about themselves, their society, their environment and their prospects.[7] Vico would have denied that language had a neutral effect on society. He realised that both primitive and modern languages carry their own particular liabilities in terms of foreshortening ethical perspective, but he maintained that this did not necessarily have to occur in either case.[8]

For Vico the sum of all the varieties of human control (unconscious more than conscious) is dominant over language. Hence language was not deterministic in his scheme; nevertheless, he would have agreed that whoever or whatever controlled language controlled society. This is why he discussed language at such great length. It was another way to get to societies, since language involved the ways in which societies looked at their lives and positions both in the world and in time.[9] It was precisely these culturally-biased assumptions which Vico felt best identified a particular society. It was because of this that he constantly argued that more attention should be paid to the social situation, both general and particular, if any true knowledge of a society were to be gained. For this reason he held that the relationship between philology and history was almost symbiotic because of a shared social structure.[10] All the while Vico desired to catch something lost, something as volatile as words.[11]

Vico was certainly more interested in the cultural nature of language than in ways in which it is distinguished from culture. He was intensely concerned with man's ability to *invent* symbols, much more than in the complexity of patterns in linguistic structure.[12] Vico declared that language was not merely one aspect of culture; it was at the very least the factor that makes possible the development, elaboration and transmission, in both oral and written form, of the accumulation of the culture as a whole.[13] Vico did not give exact answers regarding the relation of experience to language, for he believed attitudes and beliefs shaped language to a greater extent than did specific events or conditions. Therefore the relationship between the vocabulary of a society and its cultural characteristics was very close indeed.[14] He averred that philosophies and ways of life characteristic of individual cultures, even if not brought to the level of conscious formation, are reflected in their languages. For Vico the loss of the ancient form of a language would have been equivalent to the loss of the early culture of that society.

The question of the scope for originality within Vico's pre-set parameters is an issue in both languages and societies. Seemingly a discussion of the origins of language would be pointless if all the origins were the same. But, according to Vico, the absorbing aspect of the development of language or of societies was not a study of the component parts – grammar or religion, for example – but the way in which these common elements were shared to form a distinct social unit. Vico recognised the tendency to reduce a multiplicity of viewpoints to a single perspective, in order to make complicated issues simple, and he did not disapprove of it. He was convinced that these cultural universals were ideologically sound.[15] He asserted that there must be a method, a methodological justification, for historical studies. He postulated that it was necessary to discover specific analogies as a means of finding something else, for example in metaphor and metonymy. Vico's discussion of figurative speech was also important in that it recognised the importance of both the commonplace and of that which has been discarded either by the same or later societies. According to Vico, metaphorical imagination not only played a role in societies and history, but it also dealt with basic human fears and often mundane feelings.[16] Vico was not uninterested in the human ability to think or to reason. Nevertheless reason is not the new or interesting aspect of his work. The essential reality which commanded Vico's interest was social relations.

Imagination enabled Vico to examine the *mondo civile* (civil world). Although Vico examined the early stages of a civilisation at perhaps unnecessary length, the focus of his arguments was always to find the roots of an advanced society. There is no doubt that Vico was concerned with the first stages of a society exactly because they contained vital information regarding the social life of a fully functioning society.[17]

Rituals confirmed social facts, according to Vico. He viewed parallels in early society (such as primitive marriage, funeral rites and religious practices) as more than just chance. These common customs could be examined in terms of a theoretical structure, even if the historical context was not known. Marriages, burials and religious practices were not of vital interest to him simply because they were so widespread or repeated so often, so much as because they represented universally shared ways of thinking and of reacting to family life, death and the supernatural elements of the world. According to Vico, social history was a form of group remembering. He was determined to use social memory in order to reconstruct these particular social worlds. A natural development of his thought would have been an interest in what might be called social amnesia, where instances could be identified using other sources. This line of development would encompass all kinds of suppression of social memory, for suppression indicated the presence of inconvenient memories. He realised that history was written and then forgotten by the victors of the past, and he recognised that both official and unofficial memories were historical forces in their own right.[18] Vico viewed the culture of a society as indistinguishable from its collective memory. Thus myths were important for all stages of a society, because they kept the mental retrieval system of a society in good order. Myths served as the charter of the later, highly codified foundations (for example, laws) of advanced social institutions.[19]

His *degnità* were put forward without apology as the answer to the problem of reconstructing these past societies.[20] He presented his famous threes – stages of a society, social rites, customs, natural law, governments, languages, characters, jurisprudence, authority, reason – all as necessary sequences.[21] There is no evidence in Vico's writings that even he considered this model to be a preordained, predetermined pattern, much less a perfect one. The essential point in regard to Vico's cycles is that he presented them as a means to examine every aspect of society – growth, development, change, decadence, decline and dissolution. The concept of a *storia ideale eterna* encompassed every phase of the histories of all nations.[22] As a means of comparison and examination, the concepts of *corsi* and *ricorsi* were extremely important to Vico – especially as he argued that they occurred by means of divine providence, not by human intervention, and further that they had nothing to do with scholars.[23] It was not a theory of progress but of progression that he sought.

Vico did not force his methodology upon every civilisation. Rather he used these categories as a means of examination and analysis. Historical anachronisms and 'the conceit of the scholars' were, according to Vico, due to a lack of a proper historical method, which he believed had handicapped previous historians, no matter how competent.[24] These anachronisms resulted from the failure on the part of scholars to recognise the many, intertwined layers of each society's consciousness.[25] Vico's conception of philology was intertwined with his *il mondo delle nazioni* (world of nations) for philology included not only political history and the history of thought but also every other aspect of human expression. Vico was concerned with nature (from *natura* [Latin] and *nascimento* [Italian] meaning birth), not as the natural world but as the make-up, or essence, of nations. Thus an examination of the nature common to all nations was fundamental to his study of the historical development of cultures.[26] His science was composed of language, myth, literature, religion, law and economics – which together formed the basis of his new art of criticism.[27] He asserted that the truth inherent in myths had not been recognised by earlier scholars, because they possessed neither the desire nor the method necessary to discover the beliefs and assumptions (both false as well as true) which coloured all that men thought and did.[28] Scholars lacked not just the historical method but also an historical sense, an awareness of how history moved, of how the past seemed to the past.

Vico maintained that it was possible to obtain historical knowledge both of past societies and of their shared ways of thinking and feeling precisely because of these shared attitudes and prejudices. He argued that it was possible to do so because the aggregate of any social group had its own special unity which could be recreated. This was not a transcendental guiding spirit (Vico reserved that role for divine providence in all societies, and one must never forget that there is no convincing evidence that Vico was not a devout Catholic, *pace* Badaloni), rather it was the spirit, tone or personality of each particular culture which he sought. He stressed that if one is to understand man, one must think of him as a social being. This assertion alone suffices to explain Vico's strong disagreement with the Hobbesian belief in man's anti-social behaviour. For Vico, unlike Hobbes, Baruch Spinoza (1632–77) and Locke, the break between the sciences and the arts was so radical that he never considered it possible to be equally interested in both the natural elements of human development and in the social element, although he did not consider the sciences to be useless.[29]

There existed a point of genuine concurrence between Vico's views and the those of the natural law and social contract theorists; they all accepted that man needed not just a social context but a legally structured one and attempted to explain the issue of exactly how society was formed. Indeed Vico agreed with (although in his writing he only attacked) the social contract theorists in respect to his view that the initial social contact between primitive peoples was made out of fear – fear of the state of nature for Vico, fear of other men for Hobbes – and that consequently the weak were then free from the arbitrary decisions of the strong. On this point Vico and Hobbes diverge sharply. Leon Pompa sums up Vico's view very well: Vico refuted the basic tenet of the social contract theorists, for in his view society could not rest upon a contract or agreement because contacts and agreements rest upon a promise, which was dependent upon an understanding of what a promise is, which in turn only comes about via social upbringing.[30]

The natural law theorists asserted that man had certain rights which were his inalienable eternal possessions, among them the power of reason. Although a parallel might be drawn here between Vico's belief in *senso comune*, which was shared by all peoples at all time, he denied that man had natural, inalienable rights at any time, much less the power of reason at all times. For Vico man never had inalienable rights nor did man always have the faculty of reason, enabling him to tell right from wrong. Vico rejected both ideas because they failed to take into account not just the issue that man is socially conditioned but that he is *historically* conditioned.[31]

6. Historical Knowledge

Vico insisted on the uniqueness of historical knowledge and the special role of imaginative insight.[1] Imaginative knowledge, produced by the faculties of *memoria, fantasia and ingegno*, was for Vico the fundamental form of human knowledge. The *universali fantastici* functioned not only as a conception of primitive thought, but also as a principle of human knowledge itself.[2] Historical knowledge had precedence for him over all other types of knowledge, especially scientific; theological knowledge was always exempted from his discussions. It was, and to some extent still is, accepted that historical knowledge was not studied more because there was not a widespread belief in its usefulness. Vico never stated that societies learn from the past, thus it is not at all clear how the loss of knowledge regarding the past would affect a society. He had no new ideas regarding the civic virtues that children should absorb, but a modern variation of the republican virtues, the philosophy of man, expounded by Petrarch, Valla, Ficino, Pomponazzi and especially Pico can be assumed to have been Vico's model.[3] Vico's decision to omit '*La pratica*' (1731) from the last edition was perhaps because it encouraged the idea that practical application was something external to his science.[4] No ideal society was ever presented in Vico's writings either as a model or a goal because he simply was not concerned with the shaping of society in the future, but rather with analysing the social engineering of the past.[5]

Vico's entire theoretical structure was based on a belief in *senso comune* and in a *dizionario di voci mentale*, which, although elegantly simple concepts, were at the same time infinitely more sophisticated than the first words of primitive peoples. This dependence in Vico's system on some sort of instinctive inner knowledge can also be blamed for serious problems: that the human mind measures things by itself, that men ignorant of the past would judge it by the present, and the conceitedness of nations and scholars.[6]

Yet for all of the dangers inherent in reliance on inner knowledge and because of the need to avoid them, Vico maintained that the histories of obscure times could only be known if we comprehend the human nature of that time. Vico was attracted by the notion that every culture believed its views to be rational, correct and just, and perhaps what he wanted most of all was to know what *caused* such a belief to arise. Vico viewed human nature, language and society as gradually evolving entities, and held that only with a proper understanding of human nature, language or society could the temper, attitudes, prejudices and preferences of a particular period of the past be reconstructed.

Vico never regarded poetry as distinct from history. His view that poetry was an immature phase in the development of the human mind was not entirely meant as criticism of this phase. He considered the poetic mentality in a far from positive light.[7] Vico was extremely critical of early law and religion for their crudeness, falsity and failure to satisfy the demands and longings for religion, but at the same time he extolled the benefits of imagination in almost every area of early human activity.[8] This contradiction is at the very heart of his work, and if an attempt is made to explain away these differences, something intrinsic in Vico's work is lost.[9] His contempt for early societies was reserved for the so-called savage people and religions; for early poetry itself Vico had nothing but respect, almost awe.[10]

According to Ernst Cassirer (1874–1945) and Verene, sensation was a form of primitive thought for early man because only for beasts does each new sensation cancel the last.[11] But for Vico it was the attempt to reconcile the human condition to its social and physical environments through the creation of myths which was the first stage of human mental development – that the comparison of the unknown to the known, the stories, and the questioning were necessary elements even at this initial stage. There is, then, a distinction here between the view of primitive thought propounded by Cassirer and Verene, which classified instinctive reactions along with mental capacities, and that of Vico (not in spite of, but because of, his interest in imagination), who was concerned with the first manner in which human creativity was transmitted.[12] Although one could argue that in primitive societies perhaps the earliest insights were lost in the senses, now the problem is that everything is reduced to closely defined concepts and categories. For example, it is now sadly the case that far too often historical evidence is examined in isolation from the findings in other fields. Disciplinary boundaries today oppose the discovery of new knowledge, and the term interdisciplinary is often taken to imply undisciplined or vague. (Ironically, these are charges often made regarding Vico's writings, when an attempt is made to fit his ideas into an existing theoretical system.)

Vico made no such distinctions in his work. *Factum* comprised all the historical artefacts – society, law, marriage, religion and burials – of which he desired to make use.[13] It is at this point that one sees the profound significance of his belief that only by making something can we hope to understand it, and thus we can only hope to understand events in the past if we re-make them by means of our own imagination. Law was a prime example of the collective mentality, and the legal system, not surprisingly, was considered by Vico to be the crowning achievement of his age, because it was an example of human creativity uncopied from any Platonic abstract original.

Vico was intensely concerned with the construction and fashioning of the *mondo civile*. This civil world, he stated, issued from a mind often diverse, at times quite contrary, and always superior to that of the particular individuals concerned.[14] Vico considered it possible to grasp the thinking of these early societies precisely because 'the world of civil society has certainly been made by men, and that its principles [can, because they must, be rediscovered] are therefore to be found in the modifications of our same human mind' ('*che questo mondo civile egli certamente è stato fatto dagli uomini, onde se ne possono, perché se ne debbono, ritruovare i princípi dentro le modificazioni della nostra medesima mente umana.*').[15] This is the reverse of the eighteenth-century belief in preformation, a view that all physical variations of human kind were present in the first people – a type of Russian dolls effect. Vico's argument was exactly the opposite, that we hold in our subconscious minds the whole of the development of the human race, including even the most savage forms of thought and action, and that it is not only possible but mandatory that we attempt to unlock this knowledge using all methods possible – most specifically an analysis of the potentialities of 'our same human mind'. According to Berlin, these modifications were 'stages of the growth, or of the range or direction, of human thought, imagination, will, and feeling, into which any man equipped with sufficient *fantasia* (as well as rational methods) can "enter"'.[16] According to Vico, such an activity could only lead to the acquisition, not the loss, of judgement.[17]

7. *Fantasia*

Imagination had the central role in the constitution of experience and philosophy in Vico's thought. He wanted to discover the connections between forming an imaginative universal with the making of a story or myth and its physical connection (for example, fear and thunder). According to Vico it was the *universali fantastici*, which were defined by him as fables in brief, caused by fear and shame, which enabled early man to fix his attention on an object both powerful and significant.[1] These imaginative universals were indistinguishable for Vico from primitive thought,[2] and because of this he maintained that there were as many distinct cultural worlds as there were varied conditions regarding their birth and development.[3]

Vico would have disliked the term 'rigorous' used as the opposite of 'imaginative', even though he himself stated that imagination was strongest in the most fluid, least regimented stages of development in a society. He would not have accepted the implication that to be rigorous was to be rational and that this was thus the *only* stage worthy of interest. When Vico wrote that imagination decreased as reason increased, he referred to poetic creation, not to the spirit of a social group or to the recollective faculty. According to Vico, we have now lost the imaginative foundation on which understanding of the world rests.[4] This does not contradict his belief in the particular logic of each age, which was necessary for the formation of a fully functioning society. Vico's emphasis on imagination did not endanger the uniqueness of the intervening stages.[5]

Vico did not desire to return to the barbarism of early societies, but he argued that it was possible for us to examine these first social groups with our own training and prejudices and to discover admirable traits and achievements which we will never be able to duplicate. History was for Vico an ever deepening type of apprehension of the historical world.[6] Each culture has some feature not found in any other, and we cannot eliminate any of these stages of development, for each phase of a society is as essential as any other.[7]

Vico was occupied with the basic structures of the imagination. He asserted this was the way to get away from the disembodied history based on stereotypes from literary evidence which dominated in his time. Vico maintained that ideas captured through reading were generally not typical of popular culture. This was, certainly, much more true in his time than today.[8] To him the greatest advantage of non-literary sources was that, since they were not designed to inform, they were not designed to inform selectively. Modern scholars extrapolate from this premise to use buildings, monuments and the lay-out of cities as historical sources.[9]

Vico rarely discussed ancient, mediaeval or early modern views of imagination.[10] He did not discuss the standard topics, ancient or modern, of imagination – sensation, forms, fantasms or reality. Nor was he proto-Romantic in his interpretation. Imagination was for Vico best defined as creativity, not simply as images. Vico took the raw concept of imagination, without any philosophical niceties, and applied it to early societies. This is surprising, given Vico's intellectual development; it would have been more natural for him to have discussed imagination in a more technical sense. For in his time imagination was being added to the intellectual agenda, as demonstrated by many thinkers of the time (Descartes and Shaftesbury, to give two very diverse examples) who felt it necessary at least to dismiss the role of imagination.

As was discussed in Chapter 1, there was not a radical epistemological break in Vico's thought. Cartesian thought was shed in a natural and gradual process as Vico turned his attention more and more to language and societies. At no point did he become wholly anti-scientific. Nevertheless his turn toward social and historical issues was encouraged by the neo-Platonic writers that Vico absorbed during his years at Vatolla. Pico spent most of his treatise, *De imaginatione*, condemning the evils of imagination.[11] Yet, as was discussed above, it now seems clear that it was from Pico that Vico gained his fascination with the topic. That Vico had benefited greatly from earlier writers, notably Pico, Bacon and Descartes, in no way diminishes his own achievement. It was the unique synthesis of the concepts of history, society and imagination which made his work so vital.

Imagination was for Vico a true faculty in the production of images.[12] Myths and fables were used by him as the basis for the construction of primitive peoples's world view. Myths were indistinguishable to Vico from the way of thinking, the mentality, the world view of their creators. For this reason he argued that analysis of early myths and fables must be done as much as possible from the vantage points of these same peoples. He was willing to use any technique or approach to revive these lost mentalities. For example, he did not view law as radically different from literature or written histories, as has already been mentioned. This approach Vico would have seen in the neo-Platonic writers he read at Vatolla, who discussed imagination, literature and the law in the same sections of a single treatise.[13]

These themes recur throughout his writings. His earlier works listed ways in which to feed the imagination: start children in school late, where they should then read histories and study geometry, which he viewed as a necessary tool in the formation of an inventive mind.[14] Vico's discussion of the proper educational system for the children of his time becomes more theoretical when it is remembered that he compared the early Greek mentality to that of a child. Later he asserted that this was the same for all early societies.[15] Thus by discussing education, Vico was also arguing by analogy for the correct means of analysing, if not encouraging, proper development in early societies.

It should be noted that Vico occasionally discussed creative topics, in the traditional sense. In *De antiquissima italorum sapientia* in the section entitled '*De certa facultate sciendi*' ('*On the Faculty Peculiar to Knowledge*') Vico asserted that topics – the art of determining arguments and finding subject matter in rhetoric – should be considered simultaneously inventive and critical. His stress on imagination became even more marked in the later works.

The concept of memory is presented by Vico as not quite identical to or the mirror image of, imagination. In the 1744 work he wrote that memory has three aspects: memory when it remembers things, imagination when it alters or imitates them and invention when it gives them a new turn or puts them into proper arrangement and relationship.[16] In this brief passage he tied together the three concepts of *memoria, fantasia* and *ingegno*. All three stages – recollection, imitation and invention – were necessary in order for a society to pass on their cultural heritage, with the addition of their own particular contribution. Vico did not clearly distinguish between memory and imagination, because he considered that the differentiation between man's primary faculties – memory, imagination and invention – was not at all relevant to his discussion.[17] Thus although Vico identified myth as the master key to his science, it is the concept of imagination, with its recreative and recollective functions, which provides the content and explanation of his thought.

8. Other Interpretations

Vico was certainly an historicist to the extent that he considered the history of anything is a sufficient explanation of it.[1] He wrote that 'doctrines must take their beginnings from that of the matters of which they treat' ('*Le dottrine debbono cominciare da quando cominciano le materie che trattano*').[2] This does not imply that Vico accepted the complete history of a particular culture as simply the sum of its episodic parts. The essential difference is that Vico asserted the creative elements in society had the capacity to produce wildly divergent social groups from the same elements. Societies, doctrines or any social creation are moulded by their initial make-up. It is the manner in which these elements are combined that determines the final outcome.

The fundamental problem faced by some of Vico's modern interpreters is that they wish to fit Vico into the established tradition of Western philosophy.[3] The traditional philosophical approach has misinterpreted Vico in two ways. First, it does not accept that Vico considered imagination to be the prime if not the *sole* capacity involved in the construction of the human world of the past. Yet Vico clearly demonstrated that imagination was the prime capacity but not the only one. The role of rational thought in historical reconstruction was acknowledged many times by Vico, but he did not dwell long on it, because it was hardly an exciting discovery.[4] Second, it does not accept that there are areas in which it is the imagination rather than reason which is required if we are to gain such understanding of lost stages of civilisation as to recognise them as our own history.[5] Vico's stress on imagination did not imply that he meant that philosophers in the age of men must fall back on the cognitive apparatus of the previous ages. Quite the contrary, he claimed that poetic mentality was lost to scholars in the final stage of a civilisation and that imagination was the only way of retrieving this period. According to Vico, the discovery of poetic origins in some way affects our understanding of what it is to be human now.[6] A study of primitive man brings us face to face with our own ignorance and the limitations of our reason. But, unlike primitive man, we can deal with this ignorance more effectively with rational thought and our own creativity.[7]

Perhaps the main reason that Vico's concept of imagination has been so often misunderstood is that it has been taken to encompass only the faculty of producing images, of myth-making, rather than the whole range of creativity of a social group, which in turn becomes the identifying mark or badge of that people. Vico never denied the importance of the development of the human mind. He himself recognised that it was because of our advanced mental capacities that we should be able to prevent ourselves from returning to barbarism. Philosophers criticise Vico on imagination for not employing categories. But three obvious usages of imagination are clear from a careful reading of his work – imagination as the poetic faculty of early man, as the spirit of a particular social group and, finally, as the mental process we must employ in order to reconstruct these lost phases of human history.

It must be acknowledged that if one accepts Vico as a traditional philosopher in the rational tradition, the key to Vico's thought cannot be imagination. Certainly imagination has never played such a sustained role in the thinking of any other major philosopher, with the exception of Jean-Paul Sartre (1905–80), whose work was another deviation from mainline philosophy. Aristotle thought the seat of the imagination was the heart, Galen (fl. 2nd century A.D.) thought it to be the brain and Averroes took a moderate view that imagination was born in the heart and then moved to the head; imagination was nonetheless not central to pre-Vichian thought.[8]

Vico was decidedly not a conventional philosopher; his work was a radical break from traditional European thought which at most merely acknowledged the work done by earlier thinkers on imagination. Vico's thought was based on a fundamental break with the Platonic tradition, despite his enormous admiration for Plato, as well as from the much more modern scientific tradition. He strongly opposed the Enlightenment approach, which he called French, to human nature by means of an examination of the physical nature of man and his environment.[9] Vico's work was distinct from that of the *philosophes*, and it must be remembered that much of it was written in his lifetime, which sought to establish the role of the sciences and undermine the absolute authority of the church and the state.[10] Vico's theory of knowledge not only derives from but also continually relies on *fantasia*, this nonrational quality so derided by the Moderns of his time. It was this so frequently ignored or misinterpreted concept of imagination which was the key to both his thought and his new science.

9. A Critique of Berlin

Michelet's Vico was exciting but bears little resemblance to the Neapolitan thinker one reads in the original texts.[1] Croce was very good on myth but otherwise his work on Vico now seems tedious – there is far too much of

Croce's socio-political approach to life and far too little of Vico.[2] Berlin was correct to insist on the primacy of culture in Vico and to see history as a kaleidoscope of cultures. But Berlin misses several crucial topics in Vico. As his interpretation of Vico is so dominant at the moment, these omissions represent gaps in Vico studies today which must be addressed.

First, in Berlin's exultation over Vico's presentation of a theory of culture, he ignored his stress on the basic structures of society.[3] Although culture and society are very closely related concepts, the terms are not synonymous. Culture for Vico comprised the diversity of social groups; society was the shared characteristics and structure of development. One of the important reasons Vico studied early societies was that they provided a model, a basis for comparison, for more advanced social groups regarding the essential components of a society.[4] Vico desperately wanted to know exactly which elements were necessary for the composition of a society. What all three of these interpreters have failed to come to grips with is that Vico's work was concerned the *idea* of society, most especially the development of society. Michelet's Romanticism, Croce's modified Idealism and Berlin's Liberalism have taken them down very different paths, but they have somehow all missed, or considered too obvious for a prolonged examination, the notion of society. The more recent interpreters of Vico, who see Vico fundamentally as a political thinker, have a valid point in their insistence on Vico's commitment to the issue of the development of a society.[5] For Vico did indeed care passionately about the origins, make-up, composition, the very essence of societies, which are composed of human social institutions such as families, religion, law and poetry.

It was imagination which brought all these human artefacts into being, and it was imagination which distinguished one society or culture from another, differentiated by exactly how its people married, prayed and died. Vico's historical psychology of history is the dynamic element of cultural development.[6] The problem which was never solved by Vico or any of his interpreters arises when one tries to draw conclusions regarding what sort of general scheme can be concocted from a collection of mentalities.

The reason Vico desired historical knowledge and devised his plan of historical reconstruction was because he wanted to obtain specific information regarding actual societies of the past; he would then be able to compile a list of similar components of all societies. Thus his cycles, for example, are a means to these social groups, and hence to the concept of society itself. Vico cared more about the abstract concept of society than about reconstructing various past civilisations. Although presently we may be more concerned about gaining isolated glimpses of the past, every indication supports the view that Vico himself cared more about the general concept. Indeed every term in his highly idiosyncratic vocabulary leads one back to the notion of society, without which the rest of his intellectual framework has no meaning.

Second, Berlin's insistence that Vico's work was a radical break both with the Platonic tradition and with the modern scientific tradition, thus leaving no permanent values or standards, is highly useful as far as it goes. Nonetheless Berlin does not return to this point, except in passing, to discuss *senso comune* or the *dizionario di voci mentale*, the shared beliefs which do transcend all cultures, the bond between societies, without which there would be no way of comparing, much less comprehending, such disparate societies. Vico's concern with the countless configurations of universal human nature, first mentioned in the early orations, was continued in the later works in his discussions of *senso comune*.[7] According to Vico, the best means we have of ever understanding collective actions and events of the past is by examining the elements of *senso comune*. These shared attitudes are the only method left to us to discover not only the ways pre-literate societies thought and felt, but also the only means of explaining their actions, although this is much less effective. Without *senso comune* or the *dizionario di voci mentale* it would not be possible within Vico's system to make judgements regarding past societies. For some reason Berlin did not recognise this, although he very accurately pinpointed this tension in Montesquieu's work: "There is a kind of continuous dialectic in all Montesquieu's writings between absolute values which seem to correspond to the permanent interests of men as such, and those which depend upon time and place in a concrete situation."[8] Exactly the same could be said for the work of Vico, in which absolute values were called *senso comune* – slightly modified by each culture, but still distinct and clearly recognisable – and the *dizionario di voci mentale*.[9]

Third, although Berlin discusses *fantasia*, it is listed as only one of Vico's seven insights, not as the essential and singular means by which historical reconstruction of the past can proceed and historical knowledge can actually be gained.[10] The other six Vichian insights were (1) the nature of man is not static, (2) *verum-factum*, (3) the distinctive quality of historical knowledge as opposed to scientific knowledge, (4) a concept of culture and the beginning of comparative cultural anthropology, (5) creations of man are forms of self-expression and (6) as there exist no timeless Platonic principles or Forms, each human institution must be evaluated in terms of that society alone.[11] The need to examine *senso comune* and *il dizionario di voci mentale* – missing from Berlin's final point – in relation to cultural values and as the basis for critical judgements of other societies was discussed above. Therefore, it is sufficient to note here that each of these six notions is dependent on the role of *fantasia*.

Imagination as poetry was an unconscious drive in early man as the characteristic temper of a society, imagination is unavoidable. Nevertheless imagination as the means to reclaim past times, lost societies, is a deliberate effort, requiring rational thought. Vico believed it was possible by means of *fantasia* to identify individual strands of change and development in the midst of *la storia ideale eterna* which encompassed the history of all cultures.[12] It is true that one of Vico's aims was to reconstruct the histories of some of the major ancient histories known to him (in this way Vico believed he was providing both a history and a philosophy of humanity); but this was limited to references to ancient Rome. It was the method, *fantasia* – which he called philosophy – which was ultimately much more important than the examples – which he called history.[13] Vico's sophisticated usage of *fantasia* as both an unconscious and a conscious participant in historical reconstruction was not fully articulated by Berlin nor fully appreciated as the fundamental tenet of Vico's thought. Yet it was this so frequently ignored or misinterpreted concept of imagination which was the key to both Vico's thought and his new science.

Conclusion

There is at present a great deal of interest in the work of Giambattista Vico. Historians and philosophers, however, have tended to focus their attention on Vico's speculative philosophy – an ideal, eternal pattern which dictated both the development and decline of nations, to the relative neglect of his, admittedly sometimes flawed, efforts to apply his ideas regarding historical knowledge. Vico's contribution to Western thought does not depend solely upon the historical cycles or on any other rather simplistic means of reading the past backwards.[1]

Without disparaging the role of reason, Vico felt it necessary to go beyond it; he argued that only imagination could set our minds free to be able to investigate the past. Yet his strong anti-French and anti-Enlightenment views generally would have appeared reactionary to his contemporaries – if they had read him – rather than as a revision, albeit a somewhat unconscious one, of Enlightenment principles and values. It was Vico's stress on non-rational factors which separated his work from that of his contemporaries.

Desire for knowledge of the past was a given in Vico's writings. He assumed that there was a general thirst for knowledge, that knowledge of the external, natural world could never be fully obtained or mastered because it was ultimately God-made, but that historical knowledge, since it was man-made, could be grasped. Vico's philosophy of history rested on several main points.

Senso comune, 'judgement without reflection, shared by an entire class, an entire people, an entire nation, or the whole of the human race' ('*Il senso comune é un giudizio senz'alcuno riflessione, comunemente sentito da tutto un ordine, da tutto un popolo, da tutta una nazione o da tutto il gener umano*') was the vehicle by which he discussed human nature.[2] The malleability of human nature was also discussed by means of his *dizionario di voci mentale*. This was the foundation of his discussion of culture, which enabled comparison and criticism of societies very different from one's own.

Language was, according to Vico, the means by which knowledge of past culture could be obtained. Vico asserted repeatedly that language forms minds, and not minds language. He was acutely aware that language moulded the minds of later generations of a society, subconsciously teaching them the attitudes of their forefathers. It was precisely these early beliefs, attitudes, and fears – still encapsulated in later forms of a language as myths and metaphors – which contained the historical information that Vico craved. Both classical and popular mythology played a large role in Vico's attempt to unravel the patterns of history, and to Vico mythology went hand-in-hand with ritual and religion.

Vico asserted that he had found in history what he had earlier searched for in jurisprudence: eternal truths regarding historical development and historical knowledge, which transcended the mere acquisition of facts. Vico's discussion of history always had to do with distinct cultures and the idea of society. He was a social theorist in the sense that he wished to examine social groups (not those within a society, but separate cultures) and how they developed, in a similar, yet an independent and individual fashion. It is curious that, apart from a few references to pre-Homeric Greece, Vico did not put his own ideas concerning historical reconstruction into practice. Nevertheless there is much to be learned from Vico not only concerning historical knowledge but also regarding the practice of history. Within the tangled web of his ideas one finds his caveats regarding scholarly and national conceit, anachronistic tendencies and various other presumptions common when regarding the past – these were all based on Vico's belief in the creative power of the human mind.

Vico's originality lies in his historical study of cultures and society. His *scienza nuova* was *la scienza dell'uomo* (the science, study and knowledge of man). Imagination, in all its various usages, preoccupied Vico. Indeed without imagination his theoretical structure of history would be a very rough and incomplete notion. It was imagination, in the form of early poetry, which was the instrument, the means, of his study of early societies. It was imagination which embodied the spirit of a particular age of an individual culture. The faculty of imagination which produced early myths, laws or any other social institution, was identical with the faculty which created culture as a whole. Vico also had a third, distinct usage of the term: imagination which enables us *entrare* and *descendere* into the minds of these *grossi bestioni*. Gross, disgusting brutes may be the way in which he described the first men, while not diminishing his desire to comprehend their primitive, pagan and peculiar (in Vico's estimation) ways of thinking and feeling. This information was essential if any true apprehension of their society and the history of that time was ever to be gained.

Vico's writings reveal a thinker who spent his youth, once he left formal schooling and university, reading neo-Platonic writers and thus Renaissance concepts of history and imagination. As he grew increasingly disillusioned with his own personal and professional life, he abandoned even the few references in his own writings to the fashionable topics of his time – physics and mathematics, in particular. Rather than a sharp epistemological break, there was a gradual development and narrowing of his interests, to the point that by the 1730 edition of *La scienza nuova* his central themes of imagination and historical knowledge are unmistakably clear. A parallel can be drawn between Vico's own intellectual development and his discussion of the development of a society in which transitions are both natural and gradual, producing a theory of change much more sophisticated than that of Hobbes. To those familiar with his earliest works, Vico's stress on the importance of imagination and history comes as no surprise, since their outline was clear as early as in the first six orations. *Il diritto universale* is arguably more powerful (although admittedly even more obscure) than any of the versions of *La scienza nuova*. The critical point here is not an argument about texts, but a better understanding of these Vichian concepts; this can be achieved only by a familiarity with the full range of Vico's theoretical writings.

This contention that *fantasia* was Vico's central theme is one which will hopefully be of interest to those who approach Vico from the philological, literary, legal and philosophical traditions and even to those with diverse readings of Vico. One of the best arguments in support of this interpretation is that it can be effectively defended from almost any or all of his theoretical writings, including the most famous work. Not only does this view enlighten us regarding Vico's intellectual development, but it unifies to some extent the whole of his thought, by means of a common, far too often ignored, element. Nevertheless until modern editions and translations are made of the first six orations, *Il diritto universale* and the 1725 and 1730 editions of *La scienza nuova* (in particular), it is unlikely that scholars will begin to include these works in their discussions, and thus many important tenets of Vico studies, such as the validity of the epistemological break, will not be properly addressed or debated.

For Vico an historical consciousness demanded an awareness that any society under discussion was part of an ancient continuum of one society slowly becoming another, distinct and yet familiar. Thus a Vichian study of the past, particularly of one's own culture, would lead to self-awareness on the part of the historian of her own small role in the history of her society. Although Vico wrote of a changing human nature, certain elements such as *senso comune* allow investigation into these intriguing, seemingly lost cultures of the past. The corresponding development of the human mind and of society was central to his own thought, and indeed was responsible for much of the hostility towards the reception of his work in the late eighteenth century.

Vico offered a way of thinking about history, as well as the means to discover, or recover, aspects of the past. He did not proffer his theories as an infallible method, though he was not averse to calling his identification of the *storia ideale eterna* an eternal truth. Rather he made a call for caution:

Che dense notti di tenebre, che abisso di confusione non dee ingombrare e disperdere le nostre menti messe in ricerca di qual natura, di quai costumi, di qual sorta di governo dovette essere Roma antica, della quale non possiamo dalle nostre nature, costumi e governi fare nessuna, quantunque lontanissima, simiglianza! Impegnino pur i nostri ingegni tutta la loro acutezza o piú tosto arguzia per poter mantenere la riputazione alla nostra
. . . memoria, giá invecchiata in ció . . .

What dense nights of darkness must our minds encounter, in what abyss of confusion must they not be lost, as they search for the nature, the customs and the kind of government which ancient Rome must have had, unable to draw upon any likeness, no matter how remote, with our own nature, customs and governments? Let even our most ingenious [scholars] employ all their acuteness, or rather sharp-wittedness, to support the reliability of our recollections, which are indeed of great age.[3]

Vico was aware much more than some of his followers (it may not be altogether fair, but one sometimes receives the impression that the imaginative method, without empirical verification, was all that was necessary to Collingwood) of the dangers of relying only on his imaginative method, and of the importance of verifying its results furnished by every rational means available. That Vico's concept of *fantasia* has its limitations is an important realisation, for only then can one make proper use of it to obtain the historical knowledge which is sought.

Notes

CHAPTER 1: VICO'S INTELLECTUAL DEVELOPMENT

1. Vico's Orations (1699, 1700, 1701, 1704, 1705, 1707) and His Supposed Epistemological Break (1710)

1. On the widespread belief in Vico's epistemological break, see, for example, the volumes of articles compiled by the Institute for Vico Studies (New York): G. Tagliacozzo and H. White (eds), *Giambattista Vico: An International Symposium* (Baltimore: The Johns Hopkins University Press, 1969). G. Tagliacozzo, D. P. Verene (eds), *Giambattista Vico's Science of Humanity* (Baltimore: Johns Hopkins University Press, 1976). G. Tagliacozzo, M. Mooney and D. P. Verene (eds), *Vico: Past and Present* (Atlantic Highlands, New Jersey: Humanities Press, 1981). G. Tagliacozzo, M. Mooney, and D. P. Verene (eds), *Vico and Contemporary Thought* (Atlantic Highlands, New Jersey: Humanities Press, 1979, rpt. London: Macmillan, 1980) reprinted from *Social Research*, 43, Nos. 3–4 (1976).

2. Useful studies on Vico's intellectual development include N. Caraffa, *Gli studi giovanili e l'insegnamento accademico di G. B. Vico* (Urbino: Melchiorre Arduni, 1912); and N. Badaloni, *Vico prima della Scienza Nuova* (Rome: Accademia Nazionale dei Lincei, 1969).

3. *Autobiography*, pp. 3–54.

4. Anthony Kenny, *Descartes: A Study of his Philosophy* (New York: Random House, 1968) pp. 3–13. On the Cartesian influence in Italy see the following extremely useful work: B. de Besaucèle, Les Cartésians de l'Italie (Paris: Auguste Picard: 1920); also E. Garin, 'Cartesio e l'Italia', *Giornale Critico della Filosofia Italiana* 3rd series, Year 29, 4 (1950) 385–405; and F. Bouillier, *Histoire de la philosophie cartésienne*, 2 vols (Paris: Durand, 1854, 3rd edn (Paris: Delagrave, 1868).

5. Giambattista Vico, *Le Orazioni Inaugurali, Il De Italorum Sapientia e le Polemiche*, G. Gentile and F. Nicolini (eds) (Bari: Laterza, 1914) pp. 3–67. The preferred version of the Orations is now: Giambattista Vico, *Le Orazioni Inaugurali I–VI* (Bologna: Il Mulino, 1982), the first completed volume of the new edition of Vico's collected works by the Centro di Studi Vichiani. Hereafter as the Orations. See: Salvatore Monti, *Sulla tradizione e sul testo delle orazioni inaugurali di Vico* (Naples: Guida, 1977); and Maria Donzelli, *Natura e humanitas nel giovane Vico* (Naples: Istituto italiano per gli studi storici, 1970) pp. 30–68.

6. Giambattista Vico, *Il Diritto Universale*, F. Nicolini (ed.) (Bari: Laterza, 1936) 1 vol. bound in 3.

7. *Descartes: Correspondance*, C. Adam and G. Milhaud (eds), 8 vols, (Paris: Félix Alcan, Presses Universitaires de France, 1936–63). R. Descartes, *The Essential Descartes*, M. Wilson (ed.) (New York: Mentor, 1969, rpt. Scarborough, Ontario: Meridian, 1983). R. Descartes, *Oeuvres de Descartes*, C. Adam and P. Tannery (eds), 12 vols and supplement (Paris: Leopold Cerf, 1897–1913). R. Descartes, *Philosophical Works of Descartes*, E. S. Haldane and G. R. T. Ross (trs), 2 vols (Cambridge: Cambridge University Press, 1967). On *sensus communis*, see Descartes, *Discours de la méthode*; *Regulae ad directionem ingenii*, XII; and Vico, *1744*, §142, 145.

8. Descartes, *Discours de la méthode*. Vico, *Autobiography*.

9. Ibid., *Regulae ad directionem ingenii*, XII. Vico, *1744*, §1406.

10. Ibid., *Discours de la méthode*, II. Kenny, p. 4. *1744*, §330. *1725*, II, 7, Corollary, on the problems of implementing such a scheme.

11. Ibid. *Autobiography*.

12. Ibid., *Discours de la méthode*, I. See the very useful article by Yvon Belavel, 'Vico and Anti-Cartesianism', G. Tagliacozzo and H. White (eds), *Giambattista Vico: An International Symposium*, pp. 77–91.

13. Ibid., p. 77. *Autobiography*, pp. 21–22.

14. *De antiquissima italorum sapientia*. *Autobiography* (Part A, 1725).

15. Belavel, p. 77.

16. *Autobiography*. Descartes, *Discours de la méthode*, I.

17. *Autobiography*. Descartes, *Discours de la méthode*, VI.

18. Descartes, *Regulae ad directionem ingenii*, VII, VIII, XII.

19. Ibid.

20. Ibid.

21. Belavel, p. 80.

22. Descartes, *Regulae ad directionem ingenii*, III.

23. See: 'Objections V and Replies', in Haldane and Ross (trs), *The Philosophical Works of Descartes*, pp. 135–203.

24. Ibid.

25. *1744*, §34, 204–210, 400–403.

26. *De antiquissima italorum sapientia*, I.2, I.1, I.3.

27. *1744*, §349–360.

28. *De antiquissima italorum sapientia* I, II, III. C. Ciranna, *Sintesi di storia della filosofia* (Rome: Ciranna, 1986) vol. 2., pp. 138–146, esp. 139.

29. *1744*, §331.

30. Descartes, 'Objections IV, and Replies', in Haldane and Ross (trs), *The Philosophical Works of Descartes*, pp. 79–122.

31. Ibid. *1725*, I, 1.

32. Descartes, *Regulae ad directionem ingenii* (I). *1744*, §142, 1406.

33. *Autobiography*, pp. 30–31. *De nostri temporis studiorum ratione*, V.

34. Ibid. *1744*, §159.

35. On *'la scienza gli uomini'* ('on the science of man'), *1744*, §331. *Autobiography*, p. 22. See: Belavel, p. 79.

36. Ibid. *Autobiography*, p. 22.
37. Ibid., pp. 3–54.
38. *De nostri temporis studiorum ratione*, VIII.
39. *Autobiography*, pp. 3–54.
40. Donzelli. Fausto Nicolini, *La giovinezza di Giambattista Vico (1668–1700)* (Bari: Laterza, 1932).
41. Vico, *Oration I*, pp. 72–95.
42. Ibid., p. 82.
43. Ibid., pp. 72–95.
44. J. S. Mill, 'Bentham', in *Jeremy Bentham: Ten Critical Essays*, Bhikhu Parekh (ed.) (London: Frank Cass, 1974) p. 3. Rpt. from J. S. Mill, *Dissertations and Discussions*, 2 vols in 4 (London: John W. Parker and Son, 1859–1875) vol. 1, 1859, p. 333.
45. *Oration II*, pp. 96–145.
46. Ibid., p. 106.
47. *Oration III*, pp. 122–145.
48. *Oration IV*, pp. 146–165.
49. Ibid., p. 148. On the University of Naples, see Alan Ryder, *The Kingdom of Naples under Alfonso the Magnanimous: The Making of a Modern State* (Oxford: Clarendon Press, 1976) pp. 6–9.
50. *Oration V*, pp. 166–187.
51. *Oration VI*, pp. 188–209.
52. Ibid., p. 196.
53. Ibid.
54. *1744*, §283, 319.
55. Descartes, *Discours de la méthode*, I.
56. Ibid., *Regulae ad directionem ingenii*, X.
57. Ibid., *Discours de la méthode*, I.
58. Descartes, *Regulae ad directionem ingenii*, X.
59. See *1744*, §34, 204–210, 400–403.
60. *Autobiography*, pp. 3–54.
61. G. F. Pico della Mirandola, *On the Imagination*, H. Caplan (tr) (New Haven: Yale University Press, 1930).
62. See *Autobiography*, pp. 3–22.
63. Ibid.
64. I. Berlin, 'Vico and the Ideal of the Enlightenment', *Against the Current: Essays in the History of Ideas* (London: The Hogarth Press, 1979) pp. 120–129.
65. Vico, *1744* ('*La pratica*'), §1406–1407.
66. See especially *De nostri temporis studiorum ratione*.
67. *1744* ('*La pratica*'), §1406–1407.
68. Mill, p. 24 (Parekh, ed.) and p. 369 (1859 ed.). And see: R. Lefèvre, *L'Humanisme de Descartes* (Paris: Presses Universitaires de France, 1957) pp. 137–143 on Descartes and cultural development.
69. *Il diritto universale, De constantia iurisprudentis*, ch. 1. *1744*, §118, 123, 163, 338.

70. Laurence Brockliss article to be published in the forthcoming *History of European Universities*, vol. 2, by Cambridge University Press.
71. See Robert Darnton, *The Great Cat Massacre and Other Episodes in French Cultural History* (New York: Basic Books, 1984; 2nd edn, New York, Random House, 1985), ch. 5, entitled 'Philosophers Trim the Tree of Knowledge: The Epistemological Strategy of the *Encyclopédie*', pp. 191–213.

2. Verum ipsum factum

1. Donald Kunze, 'Giambattista Vico as a Philosopher of Place: Comments on a Recent Article by Mills', *Transactions of the Institute of British Geographers*, N.S. 8 (1983) 239.
2. *De antiquissima italorum sapientia*, II.
3. Vico in his reply to Article 10 of Vol. 8 of the *Giornale dei letterati d'Italia* (Venice, 1711), E. Gianturco (tr, ed.) *On the Study Methods of Our Times* (Indianapolis: Bobbs-Merrill, Library of Liberal Studies, 1965) p. xxxi. Paolo Cristofolini (ed.) *Vico: Opere Filosofiche* (Florence: Sansoni, 1971) p. 156.
4. *1744*, §331.
5. Ibid.
6. Antonio Pérez-Ramos, *Francis Bacon's Idea of Science and the Maker's Knowledge Tradition* (Oxford: Clarendon Press, 1988).
7. *De antiquissima italorum sapientia*, II, 1. *1744*, §331, 349, 376. Less controversial than his other writings on Vico but very useful is the article by James Morrison, 'Vico's Principle of *Verum* is *Factum* and the Problem of Historicism', *Journal of the History of Ideas*, 39, No. 4 (1978) 579–595. Also R. Mondolfo, *Il 'verum-factum' prima di Vico* (Naples: Guida, 1969). A. Child, *Making and Knowing in Hobbes, Vico and Dewey University of California Publications in Philosophy*, 16, No. 13 (1953). *Fare e conoscere in Hobbes, Vico e Dewey*, M. Donzelli, tr (Naples: Guida, 1970). K. Löwith, *Vicos Grundsatz: verum et factum convertuntur* (Heidelberg: Carl Winter, 1968).
8. T. Hobbes, *De cive* (Paris, 1642) [Bodleian – Seld 4 H.14 Art] and *Leviathan* (London, 1651), [Bodleian – A.1 17 Art Seld] Also Perez Zagorin's excellent article 'Vico's Theory of Knowledge: A Critique', *The Philosophical Quarterly*, 34, No. 134 (1984) 15–30, esp. 2–24.
9. Locke, *An Essay Concerning Humane [sic] Understanding* (London: Eliz. Holt for Thomas Basset, 1690). [Bodleian – LL 24 Art Seld]. Zagorin, pp. 20–24.
10. Zagorin, pp. 20–24. James Morrison, 'Vico's Principle of *Verum* is *Factum* and the Problem of Historicism', pp. 579–595.
11. Zagorin, pp. 20–24.
12. R. G. Collingwood, *The Idea of History* (Oxford: Oxford University Press, 1946, New York: Galaxy, 1956, 6th rpt.) pp. 63–66.

13. My thanks to Perez Zagorin for encouraging me to be more explicit on this point.

3. Categories of Historical Knowledge

1. Isaiah Berlin, *Vico and Herder* (London: The Hogarth Press, 1976, rpt. London: Chatto & Windus, 1980) pp. 105–114. Hereafter as Berlin.
2. Ibid., p. 108.
3. *Certum* (certainty) was often identified with authority by Vico, but this was definitely not inductive knowledge (in the Baconian sense) as Vico himself wrongly identified it. Berlin, pp. 105–106.
4. Ibid., p. 111.
5. Leon Pompa, *Vico: A Study of the 'New Science'* (Cambridge: Cambridge University Press, 1975) pp. 128–141, ch. 14, entitled 'Philosophy and Historical Confirmation'.
6. On '*la storia ideale eterna*', see, for example, *1744*, §348–349.
7. On '*knowledge per caussas*', see *1725*, I, 12. *1730*, §1291. *1744*, §345, 358, 630. On Stoic thought, *1744*, §130, 227, 335, 342, 345, 387, 585, 706. On Epicurean thought, *1744*, §5, 130, 135, 335, 338, 345, 499, 630, 696, 1109.

4. A Critique of Collingwood

1. Benedetto Croce, *La filosofia di Giambattista Vico* (Bari: Laterza, 1911). *The Philosophy of Giambattista Vico*, R. G. Collingwood (tr) (London: Howard Latimer, 1913).
2. R. G. Collingwood, *The Historical Imagination: An Inaugural Lecture 1935* (Oxford: Oxford University Press, 1935).
3. R. G. Collingwood, *The Idea of History*, on Herder, pp. 88–93, on Vico, pp. 63–71. Hereafter as Collingwood.
4. Ibid., pp. 214. See: E. Weinryb, 'Re-enactment in Retrospect', *The Monist* 72, No. 4 (1989) 568–580.
5. Collingwood, pp. 205–234, esp. 214. W. H. Walsh, *An Introduction to Philosophy of History* (London: Hutchinson University Library, 1951, 3rd edn, 1967) pp. 48–71. W. Dray, *Philosophy of History* (Englewood Cliffs, New Jersey: Prentice-Hall, 1964) pp. 10–15.
6. Patrick Gardiner, *The Nature of Historical Explanation* (Oxford: Clarendon Press, 1952, rpt. Oxford: Oxford University Press, 1968) p. 30. Paraphrased and discussed by W. Dray, *Perspectives on History* (Routledge & Kegan Paul, 1980) pp. 13, 21.
7. Dray, *Perspectives on History*, p. 16. Collingwood, pp. 205–234.
8. Collingwood, pp. 282–302. Walsh, p. 54.
9. Collingwood, pp. 282–302.
10. On *entrare e descendere*, see *1725*, II, Corollary and III, 22, 25 and *1744*, 338.

11. Louis Mink, *Mind, History and Dialectic: The Philosophy of R. G. Collingwood* (Bloomington: Indiana University Press, 1969) p. 158.

5. Historical Cycles

1. The standard view can be found as early as R. Flint, *Vico* (Edinburgh: Blackwood, 1884). More extreme views can be found in the collection, G. Tagliacozzo (ed.), *Vico and Marx: Affinities and Contrasts* (Atlantic Highlands, N.J.: Humanities Press, 1983; London: Macmillan Press, 1983).
2. See: Pietro Piovani, '*Vico senza Hegel*', in A. Corsano (ed.), *Omaggio a Vico* (Naples: Morano, 1968) pp. 551–586.
3. Ibid.
4. Berlin, p. 64.
5. *1744*, §1105–1106.
6. Collingwood, p. 67.

6. Historical Sense

1. Karl Marx, *The Eighteenth Brumaire of Louis Bonaparte* (New York: International Publishing Company, 1898) p. 5.
2. Ibid., *The Eighteenth Brumaire of Louis Bonaparte*, tr n.g. (Peking: Foreign Languages Press, 1978) p. 9. *Der achtzehnte Brumaire des Louis Bonaparte* (Hamburg, 1885) p. 1.
3. Berlin, 'Discussions on Vico', *Philosophical Quarterly*, 35, No. 140 (1985) 289. This is a response to Zagorin's article above.
4. Fausto Nicolini, *Scritti storici: Giambattista Vico* (Bari: Laterza, 1939) pp. 5–300. In 1693 Vico had written *Canzone in morte di Antonio Caraffa* and in 1719 he edited a selection of poems on the occasion of Adriano Caraffa's wedding to Tesera Borghese. See: Fausto Nicolini, *Scritti vari: Giambattista Vico*, pp. 242–243, 248–251.
5. Bruno Migliorini, *Storia della lingua italiana* (Florence: Sansoni, 1960) p. 548. Revised by T. G. Griffith as *The Italian Language* (London: Faber, 1984, 2nd English edn), ch. 10.
6. H. P. Adams, *The Life and Writings of Giambattista Vico* (London: George Allen & Unwin, 1935) p. 20.
7. Fausto Nicolini, *Giambattista Vico: Opere* (Milan, Naples: Riccardo Ricciardi, 1953) pp. 973–986.
8. Peter Burke, *Vico* (Oxford: Oxford University Press, 1985) pp. 32–34.
9. Nicolini, *Scritti storici*, Book 1, ch. 9.
10. C. E. Vaughan, *Studies in the History of Political Thought Before and After Rousseau* (Manchester: Manchester University Press, 1925) vol. 1 (of 2), pp. 237.
11. Berlin, pp. 137–139.
12. *1744*, §349. See G. Vico, '*De mente heroica*', in *Scritti vari e pagine*

sparse, F. Nicolini (ed.) (Bari: Laterza, 1940) vol. 7 of *G. B. Vico: Opere*, pp. 3–22.

13. Zagorin, pp. 24–27.
14. Ibid., 28–29. According to Zagorin Vico never clarified *who* is the knowing subject of history of the manner to which man's mind made the history it recounts (p. 28).
15. Preston King, 'Thinking Past a Problem', *The History of Ideas*, Preston King (ed.) (London: Croom Helm, 1983) p. 44.
16. Berlin, 'Discussions on Vico', p. 285.
17. Ibid., 290.
18. King, p. 55.
19. Berlin, pp. 108–109
20. King, 53.
21. Ibid., p. 22.
22. Ibid., pp. 46–49. *Il diritto universale, De constantia iurisprudentis*, I.

7. History as a Science

1. Donald R. Kelley, *The Foundations of Modern Historical Scholarship* (New York: Columbia University Press, 1970), chs. 1–2.
2. F. Bacon, *Novum organum* (1620), [Bodleian Library, Oxford – Seld 4 H 14 Art]. Zagorin, p. 27.
3. Collingwood, pp. 68–69. Collingwood's list is here expanded and modified.
4. On '*la boria delle nazioni*', see *1744*, §125–126.
5. On '*la boria de' dotti*', see *1744*, §127–128.
6. Collingwood, pp. 68–69.
7. *1744*, §1104.
8. On the 'fallacy of sources', see *1725*, II on the Twelve Tables. Collingwood, p. 69.
9. Collingwood. F. A. Wolff (1759–1824) is credited with the insight that Homer was not an actual historical person, and that the *Iliad* and the *Odyssey* were the result of the collective Greek mentality. See also K. Simonsuuri, *Homer's Original Genius: Eighteenth-Century Notions of the Early Greek Epic (1688–1798)* (Cambridge: Cambridge University Press, 1979).
10. Berlin, 'Discussions on Vico', p. 284.
11. *1744*, §149.
12. Ibid., §356.

CHAPTER TWO: VICO'S EARLY WRITINGS, 1709–28

2. *De nostri temporis studiorum ratione* (1709)

1. *De nostri temporis studiorum ratione* (1709) in *Le Orazioni Inaugurali, Il De Italorum Sapientia e le Polemiche*, G. Gentile and F. Nicolini

(eds) (Bari: Laterza, 1914), I. See also G. Vico, *On the Study Methods of Our Time*, Elio Gianturco (tr, ed.) (Indianapolis: Library of of the Liberal Arts, 1965) esp. the excellent introduction. See: Paolo Rossi, *Francesco Bacone: Dalla magica alla scienza* (Bari: Laterza, 1957) and *Le sterminate antichità: Studi Vichiana* (Pisa: Nistri-Lischi, 1969).

2. Ibid., I, X, XI.
3. G. Vico, *De mente heroica*, in F. Nicolini (ed.), *Scritti Vari e Pagine Sparse* (Bari: Laterza, 1940).
4. Ibid., *Institutiones oratoriae*, G. Crifò (ed.) (Naples: Istituto Suor Orsola Benincasa, 1989) #9, 11, 37, 38, 40, 41.
5. *De nostri temporis studiorum ratione*, XIV.
6. Ibid., III.
7. Ibid., VII.
8. Ibid.
9. Ibid., VII, VIII.
10. Ibid., VIII, III.
11. Ibid., III.
12. Ibid., II.
13. Ibid., X–XII.
14. Ibid., I–III.
15. Ibid., IV.
16. Ibid., VII.
17. Ibid.
18. Donald Davidson, 'On the Very Idea of a Conceptual Scheme', *Proceedings of the American Philosophical Association*, XLVII (1973–1974) 5–20.
19. *De nostri temporis studiorum ratione*, XII.
20. David Quint, *Origin and Originality in Renaissance Literature: Versions of the Source* (New Haven: Yale, 1983) p. x, see the preface and ch. 1 on Erasmus (1466?–1536).
21. *De nostri temporis studiorum ratione*, VIII.
22. Ibid., VII.
23. Terry Eagleton, *The Ideology of the Aesthetic* (Oxford: Basil Blackwell, 1990), ch. 2 entitled 'The Law of the Heart: Shaftesbury, Hume, Burke', pp. 31–69.

3. *De antiquissima italorum sapientia* (1710)

1. Francis Bacon, *De sapientia veterum* (1609) in *The Works of Francis Bacon*, J. Spedding, R. L. Ellis and D. D. Heath (eds) (London: Longman, 1857–1874) vol. VI, pp. 617–764.
2. G. Vico, *De antiquissima italorum sapientia* (1710) in *Le orazioni inaugurali, Il de italorum sapientia, e le polemiche*, G. Gentile and F. Nicolini (eds) (Bari: Laterza, 1914), in *G. B. Vico: Opere*, I, 1.
3. Nicolini supports this view in a footnote to the *Due risposte* in his edition

of Giambattista Vico, *Opere* (Milan, Naples: Riccardo Ricciardi, 1953) p. 310.
4. E. P. Noether, *Seeds of Italian Nationalism, 1700–1815* (New York: Columbia University Press, 1951) pp. 48–62, esp. pp. 58–59.
5. Giambattista Vico, *Opere*, F. Nicolini (ed.) (1953) p. 310.
6. *De antiquissima italorum sapientia*, ch. VII, 5.
7. Ibid.
8. Ibid., VII.3; III.
9. Ibid., VII.1.
10. Ibid., I.2, 3.
11. Ibid., V.3.
12. Ibid., VII.5.
13. Ibid., I.3.
14. Ibid., VII.4.
15. Ibid.
16. Ibid., VII,5.
17. Ibid.
18. Ibid., I,4; II,1.
19. Ibid., I, II.
20. Ibid., I,3.
21. Ibid., III.
22. Ibid.
23. Ibid., II.
24. Ibid.

4. *Il diritto universale* (1720–22)

1. *Il diritto universale, De Uno Universi Iuris Principio Et Fine Uno* in *Vico: Opere Giuridiche* (Florence: Sansoni, 1974), chs. XXXIII, LXXXII, CXLIV, CLIII, CLXX.
2. G. Vico, *De constantia philologiae* in *Il diritto universale, Parte II*, XII.25.
3. Ibid., XII.33.
4. *1725*, §272. Comma is in the original edition.
5. *Il diritto universale*, IX, XIX, XXIX–XXX, XXXII.
6. Ibid., XVIII–XIX.
7. Ibid., XV.
8. Ibid., XIII.
9. Ibid.
10. Ibid.
11. Ibid.
12. Ibid., XIV.
13. Ibid., XIV, XV.
14. Ibid., I, at the very beginning of *De constantia philologiae*.
15. Ibid. See the section in Chapter 3 on free will.
16. Ibid., XV.

17. Ibid., XVII.
18. Ibid., XI . . .
19. Ibid., XII.9.
20. Ibid., XII.
21. Ibid.
22. Ibid., XII.3.
23. Ibid., XII.11.
24. Ibid., I.5
25. Ibid., I.1, VII, XVIII.
26. Ibid., I.1.
27. Ibid., XI.

5. *La scienza nuova prima* (1725)

1. Bergin and Fisch, p. 12. *La scienza nuova in forma negativa* would be useful if only as a guide to Vico's sources.
2. Ibid., p. 16. See also Costa, 45–54.
3. G. B. Vico, *La scienza nuova*, F. Nicolini (ed.) (Bari: Laterza, 1928), 2 vols, 1113–1487, 1493–1498. See Michael Mooney, *Vico in the Tradition of Rhetoric* (Princeton: Princeton University Press, 1985) pp. 265–275 on 'Vico's Writings'.
4. H. P. Adams, *The Life and Writings of Giambattista Vico* (London: George Allen & Unwin, 1935) and T. M. Berry, *The Historical Theory of Giambattista Vico* (Washington, D.C.: Catholic University of America Press, 1949).
5. *1744*, §1405–11, especially 1406. See: *1744*, Book V on the return to barbarism.
6. G. B. Vico, *La scienza nuova prima* (1725), F. Nicolini (ed.) (Bari: Laterza, 1931), 19. Hereafter as *1725*.
7. Ibid., 41.
8. Ibid., §28.
9. Ibid., §111 and *1744*, §122, 180, 184, 198, 266.
10. Ibid., §144–6, 161–2, 240, 294, 542.
11. *1725*, §387–388
12. Ibid., §94, 96.
13. *1744*, §811, 201.
14. See *1725*, I.9, 10 and *1744*, §330 on the uselessness of texts in such a study.
15. *1725*, §100.
16. *1744*, §369–73, 37.
17. *1725*, §43.
18. *1744*, §255.
19. Ibid.
20. Ibid., §151.
21. *1725*, §210. *1744*, §376.
22. *1744*, §379.

23. *1725*, §108.
24. Ibid., §109–111.
25. *1725*, §252.
26. *Il diritto universale, De constantia iurisprudentis*, II, IV.
27. *1744*, §1104, but see also 1108.

6. Vita di Giambattista Vico scritta da sé medesimo (1725, 1728)

1. G. Costa, 'An Enduring Venetian Accomplishment: The Autobiography of G. B. Vico', *Italian Quarterly*, 21, No. 79 (1980) 46–47. See A. Battistini, *La Degnità della Retorica: Studi su G.B. Vico* (Pisa: Pacini, 1975) for an analysis of this work in terms of rhetoric.
2. G. Vico, '*Vita di Giambattista Vico scritta da sé medesimo*', *Raccolta d'opuscoli scientifici e filologici* (Venice: 1728) pp. 145–256.
3. B. Croce and F. Nicolini, *Bibliografia Vichiana* (Naples: Riccardo Ricciardi, 1947–1948) I, p. 194.
4. G. Vico, *L'Autobiografia, il carteggio e le poesie varie*, B. Croce (ed.) (Bari: Laterza, 1911) pp. 3–128.
5. G. Vico, *The Autobiography of Giambattista Vico*, M. H. Fisch and T. G. Bergin (trs) (Ithaca and London: Cornell University Press, 1944, 1983) pp. 8–19; the introduction by Fisch is first rate.
6. Gennaro Vico, *Carte vichiane donate del Marchese Villarosa per la massima parte di Gennaro Vico e a lui dirette* (Biblioteca Nazionale . . . di Napoli, XIX B43).
7. Ibid., pp. 18–19.
8. G. di Porcía, '*Progetto ai letterati d'Italia scrivere le loro vite del Signor Co(nte) Giovannartico di Porcía*', *Raccolta*, I, 128.
9. See F. Nicolini, *La giovinezza di Giambattista Vico (1668–1700)*.
10. *Autobiography*, pp. 3–22.
11. G. Vico, *Affeti di un disperato* (Naples: Philobiblon, 1948, facs. of 1693). See the introduction by Croce.
12. H. Quigley, *Italy and the Rise of a New School of Criticism in the Eighteeenth Century* (Perth: Munro & Scott, 1921), ch. 1.
13. See Giovanni Santinello, *Storia delle storie generali della filosofia* (Brescia: La Scuola, 1979) vol. II, *Dell'éta Cartesiana a Brucker*, pp. 520–635, esp. 564–632.
14. P. Giannone, *Dell'Istoria civile del Regno di Napoli* (Naples: Niccolò Naso, 1723). See also L. Marini, *Pietro Giannone e il giannonismo a Napoli nel settecento* (Bari: Laterza, 1950). G. L. C. Bedani, 'A Neglected Problem in Contemporary Vico Studies: Intellectual Freedom and Religious Constraints in Vico's Naples', *New Vico Studies*, 4 (1986) 57–72.
15. H. Acton, *The Bourbons of Naples 1734–1825* (London: Metheun, 1956), ch. 7. See also: S. Woolf, *A History of Italy* (London: Metheun, 1979) – the best book in English on the Kingdom of Naples in the eighteenth century.

16. P. C. Perrotta, '*Giambattista Vico, Philosopher-Historian*', *Catholic Historical Review*, 20 (1934–1935) 384–410. See M. Rosa, *Cattolicesismo e lumi nel settecento italiano* (Rome: Herder 1981) chs. 1–4.
17. *Autobiography*, pp. 21–22.
18. Ibid.
19. Ibid., pp. 10–11.
20. Peter Burke, *Vico* (Oxford: Oxford University Press, 1985) p. 25.

CHAPTER THREE: *LA SCIENZA NUOVA*, 1725, 1730, 1744

1. The Three Editions

1. See Chapter 2 on *La scienza nuova prima*.
2. For works on seventeenth- and eighteenth-century Naples see especially the historical works of Nicolini and N. Badaloni, *Introduzione a G. B. Vico* (Milan: Feltrinelli, 1961); and also R. Ajello (ed.), *Pietro Giannone e il suo tempo* (Naples: Jovene, 1980).
3. A. Duro (ed.) *Concordanze e indici di frequenza dei Principi di una Scienza Nuova 1725 di G. Vico* (Rome: Ateneo, 1981).
4. G. Vico, *Le orazioni inaugurali I–VI*, G. G. Visconti (ed.).
5. Granted, the Bergin and Fisch English translation was published four years before the translation of the 1744 edition (1944, 1948).
6. *Autobiography*, p. 70.
7. On the *il dizionario di voci mentale*, see: *1744*, §144–6, 161–2, 240, 294, 542, 387, 388.
8. See Nicolini edition (8 vols in 11) and Bergin and Fisch translation, *The New Science of Giambattista Vico*.
9. See the discussion of the *Correzioni, miglioramenti e aggiunte* in B. Croce and F. Nicolini, *Bibliografia Vichiana* (Naples: Laterza, 1948, 2nd ed., vol. 1 (of 2) pp. 49–53.
10. *1725*: Books 4 and 5 form less than one-fifth of the total work.
11. *1730*: pp. 1–221 of 480 pages disscuss imagination in some detail.
12. Benedetto Croce, entry on Vico in the *Encyclopaedia of the Social Sciences* (New York: Macmillan, 1935) vol. 15, pp. 249–250, esp. p. 250.
13. M. Sanna, *Catalogo vichiano napoletano* (Naples: Bibliopolis, 1986).
14. P. Cristofolini, *Opere filosofiche* (Florence: Sansoni, 1971); *Opere giurdiche* (Florence: Sansoni, 1974).

2. Vico's Vocabulary

1. On the development of human institutions: *1744*, §239–40, 249. On mental vocabularies: Ibid., §352. *1725*, III, 43.
2. *1725*, III.4–5.

3. Ibid., II.7 (Part 2, Corollary).
4. Ibid., III.6.

3. Uniformity of Ideas

1. *1744*, III.
2. *'Della discoverta del vero Dante'*, in Giambattista Vico, *Opere*, F. Nicolini (ed.) (1953) pp. 950–4.
3. *1744*, §902–4
4. Ibid., §1037.
5. *'Della discoverta del vero Dante'*, p. 950.
6. See Migliorini.
7. *1744*, §880.
8. *1725*, II.35–36; III.42.
9. *Autobiography*.
10. *1725*, I.5.
11. *1730*, §1231.
12. *1725*, II. C. E. Vaughan, *Studies in the History of Political Philosophy before and after Rousseau*, vol. 1 (of 2) pp. 204–254.
13. Ibid., II.4.
14. On innate ideas, *1725*, I.1; II.6–7.
15. *1725*, II.43.
16. On *umana mente*: *1725*, I.1.

4. Sapienza volgare

1. *1725*, II.2.
2. *1744*, §375.
3. Vico's response to the theory that language was artificial: *1725*, I.13; II.7. Further on language: *1725*, I.43.
4. *1725*, I.13.
5. Ibid., II.2.7. On religion, marriage and burials: *1744*, §333.
6. On *senso comune*: *1725*, I; and *1744*, §142.
7. But see G. Modica, *La filosofia del «senso comune» in Giambattista Vico*, (Caltanissetta-Rome: Sciascia, 1983; and his article *'Sul Ruolo del «senso comune» nel giovane Vico'*, *Rivista di filosofia neo-scolastica*, 75, No. 2 (1983) 243–262. Also J. D. Schaeffer, 'Vico's Rhetorical Model of the Mind: *Sensus communis* in the *De nostri temporis studiorum ratione'*, *Philosophy and Rhetoric* 14 (1981) 152–67.
8. On language: *1725*, I.13; II.7 and III.40–43.
9. Ibid., III.1.
10. Ibid.
11. G. Vico, *'Idea d'una grammatica filosofica'*, F. Nicolini (ed.) *Giambattista Vico: Opere* (1953) pp. 944–945..
12. On the etymologicon: *1725*, III.40–43. On the mental words, language and dictionary: *1744*: §144–6, 161–2, 240, 542, 387–388.

13. *1725*, III.43. See Berlin, *Vico and Herder*, pp. 129–30, 136.
14. C. E. Vaughan, *Studies in the History of Political Philosophy before and after Rousseau*, vol. 1 (of 2) pp. 204–254.
15. On the recognition of myths and, to a lesser degree, fables and myths as forms of cognition see: *1725*, II.7, 13; III.1. *1744*, §220–1, 375. See issue on 'Myth and Myth-Making', *Daedalus* 2 (1959).
16. *1725*, II.7, Corollary.
17. *1744*, §376. Fables as a way of thinking for an entire social group: *1744*, §816.
18. Myths as true narration: *1725*, II.
19. Poetic characters as the essence of fables: *1725*, II.5–7; *1744*, §209, 211–2, 424, 932–2, 935.
20. On *generi* or *universali fantastici*: *1744*, 34, 204–10.
21. On human utilities: *1725*, V; and *1744*, §141.
22. On Achilles: *1725*, II.4.
23. Vulgar, natural or poetic theology (theogony): *1725*, II.7. On *i poeti teologi* (theological poets): *1730*, §1296.
24. On assumptions regarding mythmaking, see David Bidney, 'Vico's *New Science of Myth*', in *Giambattista Vico: An International Symposium*, G. Tagliacozzo and H. White (eds) (Baltimore: Johns Hopkins University Press, 1969) pp. 259–277, esp. 272–274.
25. Nevertheless Vico would have agreed readily with Cassirer that the reduction of all myths to a single subject would not bring one closer to a real, that is to say, an ultimate answer. See Ernst Cassirer, *Language and Myth*, S. K. Langer (tr) (New York: Harper and Row, 1946).
26. Claude Lévi-Strauss, *Myth and Meaning* (London: Routledge & Kegan Paul, 1978), ch. 4, 'When Myth becomes History'.
27. Ibid. On mythology: *1725*, II.6–7, 13.
28. *1744*, §283, 319, 323.
29. Ibid., §122, 180, 184, 204, 319, 323, 376, 378–9. See also: *1725*, III, 1.4 . . . ; *1730*: §1181–4.
30. *1730*, §399.
31. On *curiosità*: *1725*, I.1; II.14. *1744*, §189. Negative aspect of *curiosità* according to Vico: *1725*, I.1. *1730*, II, III.
32. *1725*, §111 and *1744*, §122, 180, 184.
33. *1744*, §375.
34. Ibid., §378.
35. Ibid., §204.
36. Ibid., §283, 319.
37. Ibid., §189.

5. Religion and Society

1. References to sacred human history, for example: *1725*, I.1, 3; III.7; V.3. *1730*, §1148, 1184.
2. *1725*, I.1. *1730*, I; *1744*, 333. Religion, in Vico's writings, is always

separated from Christianity. The modern notion of the evolutionary development of religion in Vico's writings removes his distinction between the Gentile and Christian traditions – Jewish history was seldom mentioned by Vico, for the same reasons that he ignored Christian history. See especially James Morrison, 'Vico and Spinoza', *Journal of the History of Ideas*, 41, No. 1 (1980) 49–68. On this point Vico agreed with Descartes that there was a sharp distinction between philosophical and mathematical truth, on the one hand, and theological truth, on the other; for both of them this was the difference between demonstrable logical truth and revealed theological truth. Vico then, of course, went one step further to form a third category – history – which was not rational, but nonetheless comprehensible to man.

3. Berlin believes language, mythology and rites of religion to be the three most important historical artefacts. See Berlin. Also the conclusion of Donald Kelley, *The Beginning of Ideology, Consciousness and Society in the French Reformation* (Cambridge: Cambridge University Press, 1981).

4. Karl Löwith, *Meaning in History: The Theological Implications of the Philosophy of History* (Chicago: University of Chicago Press, 1949) p. 130.

5. Ibid.

6. On the social benefits of religion and marriage: *1744*, §177, 503 (religion) and 516 (marriage).

7. Warning against the return to the bestial state: *1744*, §333. See *1725*, IV, para. 5.

8. Vico on Locke, for example: *1725*, II.1; and *1730*, §1122, 1215.

9. On property: *1725*, II.15; *1744*, §531.

10. On writing, family relationships and age as issues in regard to property: *1725*. *1744*, §428, 439, 526.

11. Vico attributed the inequality of social classes to imperfect forms of religion, not to technological progress. Nevertheless this discussion has encouraged countless readers to see Vico as the prototype of Hegel and Marx. The parallels are obvious (historical cycles, social classes and some discussion of luxury by Vico), but Vico's reason for discussing these topics was to identify particular societies and thus determine the make-up of society itself. There were no transcendental or practical aspects to his work, as later ascribed to him by Croce, Gentile and Collingwood. See Giovanni Gentile, *Studi vichiani* (Florence: Felice Le Monnier, 1927, 2nd ed.).

6. Free Will

1. Human will, 'artificer of the world' and on free will: *1725*, II, III.

2. *Vici vindiciae*, X.2.

3. *1744*, §344; and also 445.

4. *1730*, §1191.

5. *1725*, II.
6. *1744*, §445.
7. Ibid., §352.
8. Ibid., §985–999, approx.
9. On *libero arbitrio*: *1725*, I, 1. On *naturali obbligazioni*: *1730*, §1269. On the shift to imagination: *1744*, §34, 204–10.
10. See Chapter 2 on the imagination and the will.

7. Formation of Society: The Taming of Primitive Man

1. *1744*, §129, 782, 1132.
2. On familial and civic authority: Ibid., §532, 524.
3. *1744*, §201.
4. *1725*, I.1. On the taming of primitive man: *1744*, §523.
5. Poetry in an educative role: *1744*, §379, 502.
6. On marriage: *1744*, §505 and especially 516.
7. *De nostri temporis studiorum ratione*, VIII.
8. Scholars: *1730*, §1140, 1161–2: *1744*, §125, most importantly 127, and 492–8 on the logic of the learned. See *De nostri temporis studiorum ratione*, III.
9. *1744*, §378, and see 399.
10. On the '*primi uomini, stupidi, insensati ed orribili*', *1744*, §374.
11. See *1744*, IV, on the various groupings of three.
12. *1725*, II.4.
13. Ibid.
14. Ibid., II.4. *1744*, §1086. See *1725*, II, §1345 on the state of the first cities.
15. *1725*, IV. See also: *1725*, III.43.
16. *1744*, 389–90.
17. Ibid.
18. Ibid., §1073–4.

8. Setti di tempi

1. See *Vico: Selected Writings*, L. Pompa (tr, ed.) (Cambridge: Cambridge University Press, 1982) p. 88, footnote 14.
2. See *1725*, I.1, 7; II.8. Also *1730*, §1384.
3. *1725*, III.43. *1744*, §443.
4. On heroes and heroic qualities: *1725*, II.17; V.8–9; *1730*, §1113. *1744*, §352, 634–61, 832, 958.
5. *1744*, V.10, and *1744*, V.
6. *1744*, §331.
7. Ibid.
8. *1730*, §1301.
9. Ibid., §1151.
10. *1725*, II.8–9.

11. Ibid., II.59.
12. Ibid., II.7.
13. Ibid., II.60.
14. See Donald R. Kelley, *The Foundations of Modern Historical Scholarship* (New York: Columbia, 1970), ch. 1 on Lorenzo Valla.
15. *1725*, I.5.
16. Ibid., I.
17. Ibid., *1744*, §394.
18. Ibid., §311.
19. See: *1725*, I.
20. *Vici Vindiciae*, VIII, XVIII.
21. James C. Morrison, 'Vico's Doctrine of the Natural Law of the Gentes', *Journal of the History of Philosophy*, 16 (1978) 47–60, esp. 52–53. *1744*, 308–309, 311.
22. Ibid., §145, 332–333. *1730*, §1163.
23. *1744*, §141.
24. Ibid., §142.
25. Ibid., §143; and *1725*, I.12.

9. New Critical Art

1. *1725*, II.9.
2. *1744*, §377 on giants and 670 on the bestial education of giants. According to Vico giants (*i giganti*, this term and also *i bestioni* were used by Vico to describe the most primitive and savage forms of man) disappeared because of the development of family structure, which was related to the fear inspired by the pagan religions. (*Il diritto universale, De constantia iurisprudentis*, XIII.)
3. *1725*, II.9.
4. *1744*, §469.
5. Ibid.
6. *1725*, IV.
7. Ibid., I.8.
8. Ibid., I.12.
9. Ibid., II.1.
10. *1730*, §1204.
11. Ibid., §1217.
12. Ibid., §1160.
13. *1744*, I.
14. *1730*, §1186.
15. *1725*, I.6.
16. *1744*, ('*La pratica*') §1405–11, most particularly 1407.
17. *1725*, I.10.
18. *1744*, §374.
19. Ibid., §499–501.
20. *1725*, II.7.
21. *1744*, §201.

22.   *1725*, I.1.
23.   *1744*, §378.
24.   *1725*, II.7, Corollary.
25.   Contrast with *1725*, IV; this is from *1744*, I, §356–8.

CHAPTER FOUR: LANGUAGE, HISTORICAL RECONSTRUCTION AND
THE DEVELOPMENT OF SOCIETY

1. Introduction

1.   The standard work is Andrea Sorrentino, *La retorica e la poetica di
     G. B. Vico* (Turin: Bocca, 1927). See also G. Bedani, 'The Poetic as
     an Aesthetic Category in Vico's *Scienza Nuova*', *Italian Studies*, 31
     (1976) 22–36. And G. Bianca, *Il concetto di poesia in Giambattista
     Vico* (Messina-Florence: G. D'Anno, 1967).
2.   G. Lepschy, 'Linguistics', in *Developing Contemporary Marxism*, Z. G.
     Barański and J. R. Short (eds) (London: Macmillan, 1985) 215. And see
     Migliorini.
3.   Croce, ch. 5.
4.   S. K. Land, 'The Account of Language in Vico's *Scienza Nuova*: A
     Critical Analysis', *Philological Quarterly* 55 (1976) 354–373.
5.   Letter from Donald Kelley, 1 September 1990.
6.   P. Burke, 'Language and anti-languages in early modern Italy', *History
     Workshop Journal*, 11 (1981) 24–32, esp. 31.
7.   *1744*, §449–454.
8.   James Burnett, *Of the Origin and Progress of Language*, 6 vols
     (Edinburgh and London: A. Kincaid and T. Cadell, 1773–92).
9.   *Dictionary of the History of Ideas* (New York: Charles Scribner's Sons,
     1973) vol. II, p. 669.
10.   J. J. Rousseau, *Discours sur l'origine e les fondamens de l'inégalité
     parmi les hommes* in *Collection complète des oeuvres de J. J. Rousseau*,
     33 vols (Geneva, 1782–89), vol. 1, pp. 50–183, Notes, pp. 184–243. *Du
     Contrat Social*, vol. 2, pp. 1–252. *Essai sur l'origine des langues*, vol. 16,
     pp. 217–325. Rousseau addressed the problem that while language
     presupposes society, the creation of human society presupposes the
     existence of language, but Vico believed the development of the two
     could not be separated. On Rousseau, see the *Dictionary of the History
     of Ideas*, vol. 2, p. 669.
11.   Burnett, 1773, vol. 1, pp. 174, 272.
12.   On the *ferini* versus the *anti-ferini* debate see: B. Labanca, *G. Vico
     e i suoi critici cattolici* (Naples: Piero, 1898); G. F. Finetti, *Difesa
     dell'autorità della sacra Scrittura contro G. Vico*, B. Croce (ed.) (Bari:
     Laterza, 1936, facs. of 1768); and *Apologia del genere umano accusato
     di essere stato una volta bestia: Parte I* (Venice: Radici, 1768).

2. Vico and Early Language

1. The standard edition of Vico's collected works is F. Nicolini, ed., with G. Gentile (vol. 1) and B. Croce (vol. 5), *G. B. Vico Opere, Scrittori d'Italia* series, 8 vols in 11 (Bari: Laterza, 1911–41).
2. This is now a contention of modern language studies. See: the introduction of E. W. Said, *The World, The Text and The Critic* (Cambridge, Mass., Harvard University Press, 1983). See also E. W. Said, *Beginnings* (New York: Basic Books, 1975) pp. 347–372 on Vico as the philosopher of beginnings.
3. This disparity is particularly obvious in the 1744 edition.
4. Nevertheless Vico was chosen to appraise Valletta's famous library indicating that Vico was recognized to have been very familiar with the libraries of Naples. See the comment by Fisch in the introduction to his translation of the Autobiography, p. 33.
5. There is a good statement of this in *1744*, §779. See also Bedani.
6. *De nostri temporis studiorum ratione*, VIII.
7. Croce (Collingwood tr.) p. 54. Laterza rpt. of Italian edition (1980), pp. 56–57.
8. Samuel Beckett began his essay entitled 'Dante . . . Bruno.Vico . . Joyce,' in *Our Exagmination Round his Factification for Incamination of Work in Progress*, in S. Beckett *et al.* (Paris: Shakespeare and Company, 1929), with a criticism of this alliance, which unfortunately was not argued more fully in the essay itself:
    > The danger is in the neatness of identifications. The conception of Philosophy and Philology as a pair of nigger minstrels out of the Teatro dei Piccoli is soothing, like the contemplation of a carefully folded ham-sandwich. Giambattista Vico himself could not resist the attractiveness of such coincidence of gesture. He insisted on complete identification between the philosophical abstraction and the empirical illustration, thereby annulling the absolutism of each conception – hoisting the real unjustifiably clear of its dimensional limits, temporalising that which is extratemporal. (p. 3)
9. *1744*, §445.
10. Ibid., §449–454. See also: L. Formigiari, 'Language and Society in the Late Eighteenth Century', *Journal of the History of Ideas* 35 (1974) 274–292.
11. Nicolini, *La giovinezza di Giambattista Vico*, p. 133. Vico wrote 'stabilimmo finalmente da ben venti anni fa, di non leggere più libri . . . ' ('I decided a good twenty years ago not to read any more books . . . ') in 1729–30.
12. Croce and Nicolini, *Bibliografia vichiana*, pp. 91–95.
13. See, especially: the *Autobiography*. R. Darnton, *The Great Cat Massacre*, p. 204.
14. Burke, p. 24.
15. Ibid., pp. 25–28.

16.   Ibid., pp. 29–30.
17.   See Max Fisch, 'Vico on Roman Law', in *Essays in Political Theory Presented to G. E. Sabine*, M. R. Konvitz and A. E. Murphy (eds) (Ithaca: Cornell University Press, 1948) pp. 62–82.
18.   The Vichian approach to mythology (although not acknowledged as such) of symbolic applications and hidden meanings has been so much in vogue in this century that Franz Boas's warning was necessary:

> As we may underestimate the poetic value of such trifling songs, we may easily overestimate the actual poetic value of stereotyped symbolic poetry that appeals to us on account of its strange imagery, but that may have to the native no other than the emotional appeal of the ritual.

See: F. Boas, *General Anthropology* (ed.) (Boston: Heath, 1938) pp. 594–595. See also R. Benedict, *Patterns of Culture* (London: Routledge, 1961, 5th rpt.).
21.   *1725*, III.
22.   S. R. Hopper, 'Myth, Dream, and Imagination' in J. Campbell, *Myths, Dreams, and Religion* (New York: E. P. Dutton, 1970) pp. 111–137, esp. p. 113.
23.   *Il diritto universale, De constantia iurisprudentia*, XII, XIX. *1744*, §443. *Autobiography*, pp. 48–54. Bedani.

3.   Language as an Instrument of Human Thought

1.   *1744*, §445.
2.   Montesquieu, *L'esprit des loix*, 2 vols (Geneva: Barrillot & Fils, 1748) vol. 1, bks. 14–17, pp. 360–443.
3.   *Aristotle's Politics*, B. Jowett (tr) (Oxford: Oxford University Press, 1938, 8th ed.) Book VIII, pp. 300–317. David Hume, 'Of National Characters' *The Philosophical Works*, T. H. Green and T. H. Grose (eds) (Darmstadt: Scientia Verlag Aalen, 1964, rpt of 1882 ed.) III, pp. 244–245.
4.   *Autobiography*, pp. 11, 25. And see *1744*, §446.
5.   I. Berlin, 'On the Pursuit of the Ideal', *New York Review of Books*, 35, No. 4 (March 17, 1988), 11–18.
6.   This argument reinforces Berlin's view of Vico against that of Momigliano presented in his searching critique of *Vico and Herder*, in which he accused both Berlin and Vico of being relativists. See A. Momigliano, 'On the Pioneer Trail', *New York Review of Books*, 23, No. 36 (November 11, 1976) 33–38.
7.   C. Ottaviano, *Metafisica dell'essere parziale* (Naples: Rondinella, 1954) pp. 585–589. Quoted in Bianca, pp. 17–18.
8.   On the metaphor, see *1744*, §404–405.
9.   G. Lakoff and M. Johnson, *Metaphors We Live By* (Chicago: University of Chicago Press, 1980).
10.   Paraphrase of M. H. Abrams by Berlin. M. H. Abrams, *The Mirror and*

the Lamp (New York: Oxford University Press, 1953) p. 285. Berlin, p. 104.

11. Berlin, p. 105.
12. *Il diritto universale, De constantia iurisprudentia*, XII.11–12.
13. Likewise issues of trade and economic relations are neglected in his theoretical works, nor is his discussion of family or socio-economic relations in the villages very sophisticated. Yet Vico wrote four works which can be termed with some accuracy as historical in nature – his autobiography, the biography of a Neapolitan general (*De rebus gestis Antonj Caraphaei*, 1716), an account of the 1701 revolt of the Neapolitan nobles (*Principum neapolitanorum coniurationis Historia*, 1703) and a recital of the War of the Spanish Succession (in *In morte di Anna Aspermont contessa di Althann*, 1724); the second and third are to be found in *Scritti storici* and the latter in *Scritti vari* (1940). Without wishing to put these four historical works on the same level as the theoretical works, it must be asserted that they are important, not only for a fuller understanding of the theoretical works but also, contrary to the accepted view, because they themselves illustrate aspects of his methods and interests which do not appear anywhere else in his writings – in particular practical economic matters.
14. M. Merleau-Ponty, *The Prose of the World*, J. O'Neill (tr) (Evanston, Ill.: Northwestern University Press, 1973) p. xiii, from 'An Unpublished Text', pp. 8–9, quoted in the Editor's Preface. *La Prose du Monde*, (Mayenne: Gallimard, 1969) pp. 7–14, 'Le fantôme d'un langage pur'.
15. *1725*, II.
16. *Il diritto universale, De constantia iurisprudentia*, XII.10.

4. The History which Vico Sought

1. *Il diritto universale, De constantia iurisprudentia*, XII.
2. *1725*, III.249
3. *Il diritto universale, De constantia iurisprudentia*, IX.91.
4. *Plato: The Collected Dialogues*, E. Hamilton and H. Cairns (eds) (Princeton: Princeton University Press, 1982, 11th pr.), *Cratylus* 414.
5. *Il diritto universale, De constantia iurisprudentia*, I.
6. Ibid., I.8. *1744*, §331. See *1725*, I.1 on '*natura comune degli uomini*' ('the common nature of men').
7. *Il diritto universale, De constantia iurisprudentia*, XII. *1744*, §34.
8. *1744*, §127–128.
10. *Il diritto universale, De constantia iurisprudentia*, I.
11. On art and music as primitive creative expressions, see Boas, ch. 6 on art (R. Benedict) and ch. 7 on literature, music and dance (F. Boas). *Il diritto universale, De constantia iurisprudentia*, I.
12. *Il diritto universale, De constantia iurisprudentia*, I.
13. E. Cassirer, *An Essay On Man: An Introduction to the Philosophy of Human Culture* (New Haven: Yale University Press, 1944) p. 94.

5. Myths

1. R. Caponigri, 'Philosophy and Philology: The "New Art of Criticism" of Giambattista Vico', *Modern Schoolman*, 59, No. 2 (1982), 81–116.
2. *1744*, §460.
3. Susanne Langer, *Philosophy in a New Key* (Cambridge, MA: Harvard University Press, 1942, 2nd edn, 15th rpt. New York: Mentor, 1951) p. 17. And see E. Sapir, *Culture, Language and Personality* (Berkeley: University of California Press, 1956).
4. W. Smalley, *Readings in Missionary Anthropology II* (South Pasadena, Ca.: William Carey Library, 1978) pp. 290–293. My thanks to Walther Olsen for bringing this book to my attention.
5. Merleau-Ponty, pp. 7–14.
6. *1744*, 352.
7. Smalley, pp. 292–299.
8. Ibid. Lévi-Strauss, pp. 15–16.
9. J. Campbell, 'Mythological Themes in Creative Art and Literature', in Campbell, *Myths, Dreams and Religions*, pp. 138–175, esp. pp. 138–144. See also: J. Campbell, *Myths We Live By* (New York: The Viking Press, 1972).
10. See Chapter 3.
11. Campbell, *Myths, Dreams and Religions*, pp. 138–144.
12. C. Lévi-Strauss, *Myth and Meaning*, p. 17. Hereafter as Lévi-Strauss.

6. Social Institutions

1. There is a danger here that the people who make the effort to read Vico (with Michelet as the classic example) often tend to believe that there is something missing in their own lives, societies or academic studies, and that these people tend thus to distort their interpretations of Vico so that he seems more out of step with his own time and the European intellectual tradition than was indeed the case.
2. *Il diritto universale, De constantia iurisprudentia*, II.
3. C. Lévi-Strauss, *Tristes tropiques* (Paris: Plon, 1955) p. 421.
4. Smalley, pp. 300–303. See M. Eliade, *Myth and Reality* (New York: Harper and Row, 1963; London: George Allen & Unwin, 1964) pp. 141–143. B. Malinowski, *Magic, Science and Religion and Other Essays* (Garden City, New York: Doubleday, 1948).
5. *1744*, §352.
6. Smalley, pp. 303–306.
7. Ibid.
8. See the introduction of Said, *The World, The Text and The Critic*.

7. Order in Myths

1. This approach is called euhemerism, a system which explains mythology as growing out of real history. See G. Cantelli, 'Myth and Language in

Vico', *Giambattista Vico's Science of Humanity*, G. Tagliacozzo and D. P. Verene (eds) (Baltimore and London: Johns Hopkins University Press, 1976) pp. 47–63. See also C. Lévi-Strauss, *La pensée sauvage* (Paris: Plon, 1962).

2. *1744*, §401–929.
3. Croce, p. 62. Croce is excellent on myth, see his ch. 5.
4. P. Rieff, *The Mind of the Moralist* (London: Victor Gollancz, 1960).
5. Smalley, pp. 310–317.
6. Lévi-Strauss, pp. 34–43.
7. Ibid.
8. Ibid., pp. 5–6. The separation between science and mythical thought in the seventeenth and eighteenth centuries was based on the works of Bacon, Newton and Descartes, for from the Renaissance to the seventeen century mythology had slipped to the background of Western thought.
9. Ibid., pp. 34–43.
10. On *senso comune*, see: 1725, I.1 and *1744*, §142.
11. Lakoff and Johnson, pp. 185–194.
12. Michel Foucault, *Histoire de la folie* (Paris: Librarie Plon, 1961); *Madness and Civilization* (New York: Random House, 1965). See the introduction of Said, *The World, The Text and The Critic*.

8. Imagination and Historical Reconstruction

1. *Il diritto universale, De constantia iurisprudentia*, I. *1744*, §34. 'To conceive his new science Vico believed it would be well to return to a state of ignorance, as if no philosophers, philologists nor books had ever existed' (Vico paraphrased by Croce and translated by Collingwood. Croce, p. 26).
2. *Il diritto universale, De constantia iurisprudentia*, I.
3. Ibid.
4. See W. Dray, *Perspectives in History*, ch. 1 on Collingwood.
5. Rieff, p. xii.
6. *1744*, §189 and Book II.
7. *1744*, §189, 184. See also G. Villa, *La filosofia del mito secondo G. B. Vico* (Milan: Fratelli Bocca, 1949).

CHAPTER FIVE: IMAGINATION AND HISTORICAL KNOWLEDGE

1. Introduction

1. *Fantasia* as expressed by poetry: *1730*, II. *1744*, §367, 375, 379, 381.
2. A notable exception was D. P. Verene's article, 'Vico's Philosophy of Imagination', Isaiah Berlin's 'Comment on Professor Verene's Paper' and Verene's 'Response by the Author', in G. Tagliacozzo, D. P. Verene (eds), *Giambattista Vico's Science of Humanity* (Baltimore:

Johns Hopkins University Press, 1976) pp. 20–36, 36–39, 39–43. In the initial article Verene drew a valuable distinction between imagination as poetry and 'imaginative recollection' – a term devised by Verene (pp. 25–27). See also his book: *Vico's Science of Imagination* (Ithaca: Cornell University Press, 1981). The various reviews of his book have served as a forum for discussion of this essential topic. See the following: A. Alberti, *The Journal of Modern History*, 55 (1983) 151–152. L. Armour, *Library Journal*, 106 (1981) 887. V. M. Bevilacqua, *Quarterly Journal of Speech*, 69 (1983) 444–447. *Bibliographical Bulletin of Philosophy*, 29 (1982) 112. A. Blasi, *Journal of the Behavioral Sciences*, 19 (1983) 265–266. S. Cain, *Religious Studies Review*, 8 (1982) 162. A. R. Caponigri, *The Modern Schoolman*, 60 (1983) 221–224. *Choice*, 19 (1981) 226. R. Dupree, *The Review of Metaphysics*, 35 (1982) 916–917. C. Evangeliou, *Philosophia*, 12 (1982) 445–447. B. A. Haddock, *Religious* Studies 19 (1983) 549–552. D. Lovekin, *Philosophy and* Rhetoric, 16 (1983) 55–60. J. Milbank, *History of European* Ideas, 4 (1983), 337–342. A. Munk, *Journal of Philosophy* and Social Science (1984) 356–357. L. Pompa, *International Studies in Philosophy*, 17, No. 1 (1985) 101–103. *Psychological Medicine*, 12 (1982). R. S. Steven, *Ethics*, 92 (1982) 792. E. F. Strong, *Journal of* the History of Philosophy, 21 (1983) 273–275. *Times Literary Supplement*, (Nov. 6, 1981) 1309. W. H. Walsh, *British Journal of Aesthetics*, 22 (1982) 378–380.

3. *Fantasia* as the spirit of a particular age: *1744*, §361–384 on poetry and the earliest stages.

4. *Fantasia* as a means of restructuring, reclaiming the past: *1744*, §349, 352–359.

5. William Dray, *Perspectives on History*, pp. 9–26. W. J. van der Dussen, *History as a Science: The Philosophy of R. G. Collingwood* (The Hague: Martinus Nijhoff, 1981).

2. *Fantasia* as the Poetic, Recreative Instinct

1 G. Lepschy, '*Fantasia e immaginazione*', *Lettere Italiane*, 39, No. 1 (1987) 20–34. And see: Luigi Ambrosi, *La Psicologia dell'imaginazione nella storia della filosofia* (Rome: Dante Aligheri, 1898, rpt. Padua: CEDAM, 1959).

2. *De nostri temporis studiorum ratione*, III.

3. *De antiquissima italorum sapientia*, V, 3; VII, 4.

4. *De nostri temporis studiorum ratione*, III. On geometry: *De antiquissima italorum sapientia*, I.2; II.1; III; IV.2; VII.

5. *Il diritto universale*, *De constantia iurisprudentia*, XIV–XV. *De antiquissima italorum sapientia*, I.2, II.1, III, IV.2, VII.

6. Ibid., VII.1, I.

7. *Il diritto universale*, *De constantia iurisprudentia*, XII.25, XIII.

8. Ibid.

9. *De antiquissima italorum sapientia*, III.
10. Ibid., VII.5. *1744*, §122.
11. See Chapter 2 on *De nostri temporis studiorum ratione* and *La scienza nuova prima*.
12. *1730*, §1186.
13. *Il diritto universale, De constantia iurisprudentia*, XII.2.
14. *De nostri temporis studiorum ratione*, II.
15. Ibid., III. *1725*, II.9, 16. *1744*, §699.
16. *1744*, §122, 180, 189.
17. Ibid., §189.
18. Ibid., §461–462.
19. Vico attacked various authorities on language in Book II of the 1744 edition entitled '*Della sapienza poetica*'.
20. Ibid.
21. Cecil Sprigge, *Benedetto Croce: Man and Thinker* (Cambridge: Bowes and Bowes, 1952) ch. 2.

3. *Fantasia* as the Expression of the Spirit of a Particular Age

1. *1744*, §361–384.
2. Berlin, p. 138.
3. J. C. Robertson, *Studies in the Genesis of the Romantic Theory in the Eighteenth Century* (Cambridge: Cambridge University Press, 1923) pp. 179–194, ch. VIII.
4. *De antiquissima italorum sapientia*, VII. *1730, 1744* II.
5. van der Dussen, 2.5.

4. *Fantasia* as *la scienza nuova*

1. *De antiquissima italorum sapientia*, VII. *1730* and *1744*, II.
2. Berlin, pp. xvi on human nature. On *senso comune*, see: *De nostri temporis studiorum ratione*, III. *De antiquissima italorum sapientia*, VII.5. *1744*, §142.
3. *On the Study Methods of Our Time*, E. Gianturco (tr, ed.) p. xvi.
4. *1744*, I.
5. On *umana mente*, see *1725*, I.1 and *1744*, §331.
6. *1744*, I, section 4, §338–360.
7. On the dictionary of mental words: *1725*, III.43. *1744*, §144–146, 161–162, 240, 294, 542.
8. *1744*, §337.
9. The titles of all three versions of *La scienza nuova* are as follows: *Princípi di una scienza nuova intorno alla natura delle nazioni per la quale si ritruovano i princípi di altro sistema del diritto naturale delle genti* (1725). *Cinque libri di G. B. Vico de' Principi di una Scienza Nuova d'intorno alla comune natura delle Nazioni, in questa seconda impression con più propia maniera condotti, e di molto accresciuti*

(1730). *Principj di scienza nuova di Giambattista Vico d'intorno alla comune natura delle nazioni* (1744).
10. On *pudore*, *1725*, I.1.
11. *1744*, §125–128.
12. *De nostri temporis studiorum ratione*, III.
13. *1744*, §125–128.
14. Ibid., §125–126.
15. Ibid., §127–128.

5. History

1. Vico very often dealt with the gradual but creative force of imagination; for example, *1744*, §147: 'The nature of institutions is nothing but their coming into being *nascimento*) at certain times and in certain guises. whenever the time and guise are thus and so, such and not otherwise are the institutions that come into being'. See *1744*, §148–149. History was for Vico 'the testimony of the times' (*Il diritto universale, De constantia iurisprudentia* II.5.
2. This is the title of Book V of *1730* and *1744*.
3. *De antiquissima italorum sapientia*, VII. 4. Pompa, *Vico: A Study of* the 'New Science', pp. 77–78.
4. On the cultural unconscious, see Carlo Ginzburg, *The Cheese and the Worms: The Cosmos of a Sixteenth-Century Miller*, John and Anne Tedeschi (trs) (Harmondsworth: Penguin, 1987, rpt.).
5. *De antiquissima italorum sapientia*, II.
6. Carlo Ginzburg and Peter Burke's discussion, Institute of Contemporary Arts, London, 15 October 1987.
7. *1730* and *1744*, II.
8. Milbank review of Verene, p. 339.
9. *1730* and *1744*, II.
10. *Il diritto universale, De constantia iurisprudentia*, I.1, VII on philology and history. Milbank's review of Verene, p. 339.
11. Ginzburg (15 October 1987).
12. *International Encyclopedia of the Social Sciences* (New York: Macmillan, 1968), vol. 9, p. 19.
13. Ibid., p. 20.
14. Ginzburg (15 October 1987).
15. *1730* and *1744*, II. Bianca, ch. 3.
16. *1744*, §349.
17. *1744*, §354–358.
18. Peter Burke, 'History as social memory', paper delivered at Wolfson College, Oxford, 16 February 1988, in the College Lecture series on Memory.
19. Thomas Butler, 'Memory: a mixed blessing', paper delivered at Wolfson College, Oxford, 19 January 1988, in the College Lecture series on Memory.

20. As early as *Il diritto universale, De constantia iurisprudentia,* (XII.3) in 1721, Vico claimed his *degnità* as the necessary basis of any discussion of (the causes of the ignorance surrounding) the origin of poetry.
21. *1744,* V.
22. Vico went so far as to state in the 1730 edition that the humanity of the nations, like everything else that is mortal, must run and finish its course. Caponigri article. T. Berry, *The Historical Theory of Giambattista Vico* (Washington, D.C.: Catholic University of America, 1949).
23. A. R. Caponigri, *Time and Idea: The Theory of History in Giambattista Vico* (London: Routledge and Kegan Paul, 1953) pp. 133, 136.
24. Bruce Haddock, 'Vico and the Methodology of the History of Ideas', G. Tagliacozzo, D. P. Verene (eds), *Vico: Past and Present* (Atlantic Highlands, New Jersey: Humanities Press International, 1981) 227–239, esp. 238.
25. Ibid., pp. 230–231.
26. On '*la natura comune delle nazioni*': *1744,* §412.
27. E. Auerbach, *Literary Language and Its Public in Late Latin Antiquity and in the Middle Ages* (London: Routledge & Kegan Paul, 1965) pp. 5–24.
28. Ibid., p. 231.
29. L. Pompa, 'Vico's Theory of the Causes of Historical Change', (Tunbridge Wells, Kent: Institute for Cultural Research, 1971, 1979) pp. 6–7.
30. Ibid., pp. 6–8. Hobbes, *Leviathan.*
31. Ibid. H. Grotius, *De Jure Belli Ac Pacis Libri Tres,* F. W. Kelsey (tr), 2 vols (Oxford: Clarendon Press, 1925). Aquinas, *Selected Political Writings,* A. P. d'Entrèves (ed.) and J. G. Dawson (tr) (Oxford: Oxford University Press, 1948).

6. Historical Knowledge

1. James Morrison, 'Three Interpretations of Vico', *Journal of the History of Ideas,* 39, No. 3 (1978) 515.
2. *1744,* §34, 204, 210, 400–403. See the review of Verene by A. Alberti, *The Journal of Modern History,* 55 (1983), 151–152, esp. 151.
3. *De nostri temporis studiorum ratione. 'La Pratica'.* J. Habermas, *Theory and Practice,* J. Viertel (tr) (London: Heinemann, 1974) pp. 45–46, 79. Habermas viewed *La scienza nuova* as the transition from civic humanism to modern social philosophy. See the Milbank review of Verene.
4. *1744,* §428, 452. Jeffrey Barnouw's review of *Vico's Science of Humanity, Eighteenth Century Studies,* 10 (1977) 388.
5. See Chapter 3.
6. *1744,* §120–121, 125–128.
7. Pompa was exactly right that 'while he (Vico) praises the imaginative products of early man for their incomparable beauty and immediacy, he is equally critical of early law and early religion for their crudeness,

falsity, and failure to satisfy the demands of religion'. Review of Verene by Leon Pompa in *International Studies in Philosophy*, 17, No. 1 (1985) 101–103, esp. 102. But Pompa was not correct (in my view) to then diminish the importance of imagination in Vico's thought because of this ambivalence of Vico's towards early societies.

8. On imagination, see *1730, 1744*, II.
9. Gardiner's review of Berlin, *History and Theory*, 16, No. 1 (1977), 45–51, esp. 45.
10. Bianca. Milbank's review of Verene, pp. 337–342.
11. Strong's review of Verene, pp. 273–275.
12. This idea was stated again and again by Vico: for example, *1744*, §360.
13. There are 70 specific references to the Law of the Twelve Tables in the 1744 edition alone (there were much longer passages in the 1725 edition). Vico postulated that the Twelve Tables represented the early wisdom of the ancient Italian peoples eventually put in a codified version. He used this same argument in '*La Discoverta del vero Omero*' (*1730, 1744*, III).
14. *1744*, §1108.
15. *1744*, §331, Bergin and Fisch translation. Pompa translation in brackets: *Selected Writings*, L. Pompa (tr, ed.) pp. 2016. On preformation, see the *Dictionary of the History of Ideas*, vol. 1, p. 231; vol. 2., p. 178, 284; vol. 4, pp. 309–310.
16. *1744*, §349. Berlin, p. 27.
17. E. Auerbach, *Literary Language and Its Public in Late Latin Antiquity and in the Middle Ages*, p. 19.

7. *Fantasia*

1. *1744*, §403.
2. *1744*, §495–8.
3. *1725*, II.9.
4. Joseph Campbell, *Myths to live by*, pp. 10–20.
5. Excellent review of Verene by W. H. Walsh, *British Journal of Aesthetics*, 22 (1982) 378–380.
6. Berlin.
7. Berlin, 'On the Pursuit of the Ideal'.
8. *1744*, §384, 330.
9. Robert Darnton, *The Great Cat Massacre and Other Episodes in French Cultural History*, pp. 107–143, on the lay-out of cities as historical sources.
10. *1744*, §699, 819. G. S. Brett, *A History of Psychology*, 3 vols (London: George Allen & Unwin, 1912, 1921). W. Bundy, *The Theory of Imagination in Classical and Medieval Thought*, University of Illinois Studies in Language and Literature, 12, Nos. 2–3 (1927) 1–289.
11. Pico della Mirandola, *On the Imagination*, H. Caplan, tr.

12.   *De antiquissima italorum sapientia*, VII.1.
13. *Autobiography*, pp. 3–54.
14. *1744*, §496.
15. Milbank's review of Verene, p. 338.
16   *1744*, §819.
17. See J. M. Shorter, 'Imagination', *Mind*, N.S. 61 (1952) 528–542.

8. Other Interpretations

1.   D. E. Lee and R. N. Beck, 'The Meaning of "Historicism"', *American Historical Review*, 59, No. 3 (1959) 568–577, esp. 568. And D. D. Runes (ed), *The Dictionary of Philosophy* (London: George Routledge & Sons, 1944) p. 127.
2.   *1744*, §314.
3.   Pompa, *Vico: A Study of the 'New Science'* and his review of Verene.
4.   Ibid.
5.   Ibid.
6.   *1744*, §378–380.
7.   Milbank's review of Verene, p. 340.
8.   Pico, pp. 34–37.
9.   See: *On the Study Methods of Our Times*, Gianturco (tr, ed.), pp. ix–xxxiii of the introduction.
10. For two of the best-known works on the Enlightenment see P. Gay, *The Enlightenment: An Interpretation*, 2 vols (London: Weidenfeld and Nicolson, 1966); and P. Hazard, *La Crise de la conscience européenne, 1680–1715* (Paris: Boivin, 1935). *The European Mind*, J. L. May (tr) (Harmondsworth: Penguin, 1973, 3rd edn).

9. A Critique of Berlin

1.   J. B. Vico, *Principes de la philosophie de l'histoire*, J. Michelet, (tr, ed.) (Paris: Jules Renouard, 1827, rpt. 1859). J. B. Vico, *Oeuvres choisies de Vico*, J. Michelet (tr, ed.) (Paris: Ernest Flammarion, 1895 [?]).
2.   Croce.
3.   Berlin, pp. xvi–xix, 109–110.
4.   *1744*, §152, 247, 662.
5.   G. Giarrizzo, *Vico: La politica e la storia* (Naples: Guida, 1981). B. A. Haddock, *Vico's Political Thought* (Swansea: Mortlake Press, 1986). F. Vaughan, *The Political Philosophy of Giambattista Vico* (The Hague: Martinus Nijhoff, 1972).
6.   *1744*, §384.
7.   On *senso commune*, *De antiquissima italorum sapientia*, 5. *1725*, I, 1. See J. Morrison 'Three Interpretations of Vico', p. 514.
8.   Berlin, article on 'Montesquieu' in *Against the Current*, pp. 130–161, 157.
9.   See: Momigliano's attack on Berlin and Vico, accusing them both of

of relativism, and Berlin's response, 'Note on Alleged Relativism in Eighteenth Century Thought' in *Substance and Form in History*, Leon Pompa and W. Dray (eds) (Edinburgh: Edinburgh University Press, 1981) pp. 1–14.

10. Berlin, pp. xvi–xx.
11. Ibid.
12. *1744*, §393 and the last section of Book V.
13. *1744*, §346, 148. Morrison, 'Three Interpretations of Vico', pp. 511–518. On a perfect society, see G. Huppert, *The Idea of Perfect History* (Urbana: University of Illinois Press, 1970).

CONCLUSION

1. King, p. 55.
2. *1744*, §142.
3. *1725*, II.7, Corollary.

Bibliography

1. Manuscripts and First Editions

The first two sections of this list follow the model of M. Sanna, *Catalogo vichiano napoletano* (Naples: Bibliopolis, 1986).

Titles, descriptions and shelfmarks are, if not indicated otherwise, those used in the Manuscript and Rare Books Room of the *Biblioteca Nazionale «Vittorio Emanuele III» di Napoli*.

a. Manuscripts

Sei fascicoli di carte vichiane varie non rilegate (XIX 42)
  Versi iscrizioni del Vico e al Vico
  Frammenti di scritti vari del Vico
  Lettere del Vico e al Vico o riguardanti Vico
  Carte varie della scuola del Vico
  Un'opera per commissione; Ragionamento primo e secondo
  Carte varie relative alla vita e alla fortuna del Vico

Scienza nuova ed altri scritti autografi (XIII-D 79–80)
  Scienza nuova
  Correzioni, miglioramenti e aggiunte terze poste insieme con le prime della
    Scienza Nuova Seconda
  Vici Vindiciae
  Oratio «Si umquam Divina Providentia . . . .»

Autografo cartaceo (in miscellanea) (XIII H 50)

Codice cartaceo (XIII B 55)
  Sommario autografo
  Autografo dell Orazione inaugurali
  Autografo delle Emendationes alle Orazioni

Codice cartaceo (XVII B 30)
  Autografo di Correzioni, miglioramenti ed aggiunte Terze poste insieme
    con le Prime e Seconde e tutte coordinate per incorporarsi nella Terza
    impressione della Scienza Nuova
  Un'orazione latine autografa per le nozze di Carlo di Borbone «Si umquam
    Divina Providentia . .»
  Iscrizione per Jacope Stuart

Apografo di una lettera a D. Francesco.

Ms. *cartaceo* (XIII B73)
   Delle cene sontuose de' romani

Cartaceo apografo (transcriptions) (XVIII 38)
   De chriss
   Collectio phrasium elocutionum ac rerum notabilium selectae . . .
   In artem poeticam Q. Horatii Flacci
   Le sei commedie di Terenzio tradotte da G. B. Vico (1735)

G(ennaro) Vico, *Carte vichiane donate dal Marchese di Villarosa per la massima parte di Gennaro Vico e a lui dirette* (XIX B 43)

b. First editions
De antiquissima italorum sapientia ex linguae latinae originibus eruenda (Naples: Felice Mosca, 1710). [Houghton Library, Harvard – *IC7. V6643.710d].

De nostri temporis studiorum ratione (Naples: Felice Mosca, 1709). [Bodleian Library, Oxford – 8 St. Amand 74]. [Houghton Library, Harvard – *IC7. V6643.709d].

Cinque libri di G. B. Vico de' Principi di una Scienza Nuova d'intorno alla comune nature delle Nazioni, in questa seconda impressione con più propia maniera condotti, e di molto accresciuti (Naples: Felice Mosca, 1730). (XVIII 39)

Fogli volanti (XIV G 20 [1])
   Vico, G. *Spese matrimoniali,* in *Raccolta (grande) di cose varie ecc.*

Libri a stampa con note autografe (XIII B 62)
   De Universi Juris uno principio, et fine Uno (Naples: Felice Mosca, 1720).
   [Law Library, Special Collections, Harvard].
   De constantia jurisprudentis (Naples: Felice Mosca, 1721).
   Notae in duos libros alterum De uno . . . *alterum De constanti* . . . (Naples:
   Felice Mosca, 1722).
   Sinopsi del Diritto universale (Naples: Felice Mosca 1720).
   Mendorum ab typis literarris emendationes.

Princìpi di una scienza nuova (Naples: Felice Mosca, 1725). [Houghton Library, Harvard – *C7. V6643.725p, with corrections in Vico's hand].

Principii d'una scienza nuova (Naples: Felice Mosca, 1730). (XIII H 58 and XIII H 59, 2 copies)

Princìpi di una scienza nuova (Naples: Felice Mosca, 1744). [Taylor Institute Library, Oxford – Moore 4 f.3]. [Houghton Library, Harvard – *IC7.V6643.725pc].

De Rebus Gesti Antonj Caraphaei (Naples: Felice Mosca, 1716). [Houghton Library, Harvard – Typ 725. 16.869]

2. Other Editions of Vico's Writings

Latinae Orationes (Naples: Josephus Raymundus, 1766). [Houghton Library, Harvard – *IC7.V6643.766l]
Principes de la philosophie de l'histoire, J. Michelet (ed., tr) (Paris: Jules Renouard, 1826, rpt. 1859).
Opere di Giambattista Vico, G. Ferrari (ed.), 2nd edn, 6 vols (Milan: Società tipografica de' classici italiani, 1852–1854).
Opere di Giambattista Vico, F. Pomodoro (ed.), 8 vols (Naples: Morano, previously *Tipografia de' classici latini*, 1858–1869, rpt. 8 vols in 4, Leipzig: *Zentralantiquariat der Deutschen Demokratischen Republik*, 1970).
Oeuvres choisies de Vico, J. Michelet (ed., tr) (Paris: Ernest Flammarion, 1895[?]).
Opere di G. B. Vico, F. Nicolini (ed.) with G. Gentile (vol. 1) and B. Croce (vol. 5), (eds), 8 vols in 11 (Bari: Laterza, 1911–1941).
La Scienza nuova e opere scelte di G. B. Vico, N. Abbagnano (ed.) (Turin: Unione tipografico-editrice torinese, 1952).
Opere di Giambattista Vico, F. Nicolini (ed.) (Milan and Naples: Ricciardi, 1953).
Tutte le opere di Giambattista Vico, F. Flora (ed.) (Milan: Mondadori, 1957–) 1 vol. to date.
Opere di Giambattista Vico, P. Rossi (ed.) (Milan: Rizzoli, 1959).
Giambattista Vico: Opere filosofiche, P. Cristofolini (ed.) (Florence: Sansoni, 1971).
Giambattista Vico: Opere giuridiche, P. Cristofolini (ed.) (Florence: Sansoni, 1974).
Le Orazioni Inaugurali I-VI, G. G. Visconti (ed.) (Bolgona: Il Mulino, 1982) vol. 1 of the new edition of Vico by the Centro di Studi Vichiani.

a. A select list of later editions of single or several works by Vico
La conguira dei principi napoletani del 1701, E. De Falco (ed.) (Naples: Istituto Editoriale del Mezzogiorno, 1971).
Institutiones Oratoriae, G. Crifò (ed.) (Naples: Istituto Suro Orsola Benincasa, 1989).
De nostri temporis studiorum ratione, P. Massimi (ed.) (Rome: Armando Armando, 1974).
Opere, R. Parenti (ed.), 2 vols (Naples: Fulvio Rossi, 1972).
Principj di una scienza nuova intorno alla natura delle nazioni, T. Gregory

(ed.) (Rome: Ateno, 1979, facs. of 1725 edn) vol. 1. *Concordanze e indici di frequenza dei Principi di una Scienza Nuova 1725 di G. Vico* (Rome: Ateno, 1981), vol. 2.
Principes d'une science nouvelle relative a la nature commune des nations (Paris: Nagel, 1953).
La scienza nuova, Paolo Rossi (ed.) (Milan: Rizzoli, 1977).

b. English translations of Vico's works
The Autobiography of Giambattista Vico, M. H. Fisch and T. G. Bergin (trs, eds) (Ithaca: Cornell University Press, 1944, 1983).
The New Science of Giambattista Vico, M. H. Fisch and T. G. Bergin (trs, eds) (Ithaca: Cornell University Press, 1948, 1968).
On the Study Methods of Our Time, E. Gianturco (tr, ed.) (Indianapolis, Indiana: Bobbs-Merrill, Library of the Liberal Arts, 1965).
'On the Heroic Mind', E. Sewell and A. C. Sirignano (trs), *Social Research*, 43 (1976): 886–903. Rpt. *Vico and Contemporary Thought*, G. Tagliacozzo, M. Mooney, D. P. Verene (eds) (Atlantic Highlands, New Jersey: Humanities Press, 1979) 228–245.
Vico: Selected Writings, L. Pompa (tr, ed.) (Cambridge: Cambridge University Press, 1982).

3. Other Primary Sources

Bacon, F., *Novum Organum* (1620). [Bodleian – Byw. C 1. 24].
Burnett, J., *Of the Origin and Progress of Language*, 6 vols (Edinburgh, London: A. Kincaid and T. Cadell, 1773–92). [Bodleian Library, Oxford – 8 Q 25–30 Linc.].
Doria, P. M., *La vita civile con un trattato della educazione del Principe* (Naples, 1710). [Bodleian – Vet. D4 d.60].
Giannone, P., *Dell'istoria civile del Regno di Napoli* (Naples: Niccolò Naso, 1723). [Bodleian Library, Oxford – Vet. F4 d. 7–10] *The Civil History of the Kingdom of Naples*, James Ogilvie (tr) (London, 1729). [Bodleian – Vet A4 c 231, 232].
Hobbes, T., *De cive* (Paris, 1642). [Bodleian Library, Oxford - Seld 4 H.14 Art].
Hobbes, T., *Leviathan* (London, 1651). [Bodleian Library, Oxford - A.1 17 Art Seld].
Le Clerc, J., *Bibliothèque ancienne et moderne*, 20 vols bound in 11 (Amsterdam: Frères Wetstein, 1714–23). [Taylor Institution Library, Oxford – K 43–45].
Locke, J., *An Essay Concerning Humane [sic] Understanding* (London: Eliz. Holt for Thomas Basset, 1690). [Bodleian Library, Oxford – LL 24 Art Seld].
Michelet, *Introduction a l'Histoire Universelle* (Paris: Hachette, 1843, 3rd edn). [Taylor Institution Library, Oxford – Vet Fr. III B 445].

Montesquieu, *L'Esprit des loix [sic]*, 2 vols (Geneva: Barillot & Fils, 1748).
[Bodleian Library, Oxford – EE 133, 134 Art].
Muratori, L., *Della forza della fantasia umana* (Venice: G. Pasquali, 1745, rpt
Venice: Alvisopoli, 1825). [Biblioteca Nazionale . . . di Napoli].
Rousseau, J. J., *Collection complete des oeuvres de J. J. Rousseau*, 33 vols
(Geneva, 1782–89). [Taylor Institution Library, Oxford – VR.1 1782–89].

4. Secondary Works

Articles from the *Bolletino del Centro di Studi Vichiani, New Vico Studies*,
the volumes produced by the Institute for Vico Studies (New York) and
for the tricentenary of Vico's birth (1968) are not listed separately in the
bibliography.

Abbagnano, N., 'Vico e l'Illuminismo: risposta a F. Nicolini', *Rivista di
Filosofia* 14, No. 3 (1953) 338–342.
Abrams, M. H., *The Mirror and the Lamp* (New York: Oxford University Press,
1953).
Acton, H., *The Bourbons of Naples 1734–1825* (London: Metheun, 1956).
Adams, H. P., *The Life and Writings of Giambattista Vico* (London: George
Allen & Unwin, 1935, rpt. New York: Russell & Russell, 1970).
Agrimi, M., 'Vico Oggi', *itinerai*, Nos. 1–2 (1981).
Ajello, R., *Pietro Giannone e il suo tempo*, 2 vols (Naples: Jovene, 1980).
Alatri, P., '*Un Convegno su Illuministi e Giacobini a Napoli*', *Studi Storici*, 23,
No. 2 (1982) 439–448.
Alberti, A., 'Primitive Language and Feudal Ideology: A Discovery of Vico',
European Institute Colloquium Papers (28–30 September 1983).
Alston, W. P., *Philosophy of Language* (Englewood Cliffs, New Jersey:
Prentice-Hall, 1964).
Amerio, F., *Introduzione allo studio di G. B. Vico* (Torino: Società Editrice
Internazionale, 1947).
Ambrosi, L., *La Psicologia dell'imaginazione nella storia della filosofia*
(Rome: Dante Aligheri, 1898, rpt. Padua: CEDAM, 1959).
Anderson, M. S., *Historians and Eighteenth Century Europe 1715–89* (Oxford:
Clarendon Press, 1979).
Aristotle, *De Anima (On the Soul)*, H. Lawson-Tancred (tr) (Harmondsworth:
Penguin, 1986).
Aristotle's Politics, B. Jowett (ed., tr) (Oxford: Oxford University Press, 1938,
8th edn).
Art and Ideas in Eighteenth Century Italy (Rome: Edizioni di Storia e
Letteratura, 1960).
Aquinas, *Selected Political Writings*, A. P. d'Entrèves (ed.) and J. G. Dawson
(tr) (Oxford: Oxford University Press, 1948).
Auerbach, E., *Literary Language and Its Public in Late Latin Antiquity and in
the Middle Ages* (London: Routledge & Kegan Paul, 1965).

Auerbach, E., 'Vico and Aesthetic Historicism', *The Journal of Aesthetics and Art Criticism*, 8 (1949), 110–118, rpt. in *Scenes from the Drama of European Literature*, New York: Meridian, 1969).

Bacon, F., *The Advancement of Learning and New Atlantis*, A. Johnston (ed.) (Oxford: Clarendon Press, 1974).

Bacon, F., *The Works of Francis Bacon*, J. Spedding, R. L. Ellis and D. D. Heath (eds) (London: Longman, 1857–1874).

Badaloni, N., *Antonio Conti: Un abate libero pensatore tra Newton e Voltaire* (Milan: Feltrinelli, 1968).

Badaloni, N., *Introduzione a G. B. Vico* (Milan: Feltrinelli, 1961).

Badaloni, N., *Vico prima della Scienza Nuova* (Rome: Accademia Nazionale dei Lincei, 1969).

Baker, J. V., *The Sacred River: Coleridge's Theory of the Imagination* (Baton Rouge: Louisiana State University Press, 1957).

Barański, Z. G. and J. R. Short (eds), *Developing Contemporary Marxism* (London: Macmillan, 1985).

Barbagallo, C., 'The Conditions and Tendencies in Historical Writing in Italy Today', *Journal of Modern History*, 1, No. 2 (1929) 236–244.

Barber, W. H. *et al.*, (eds), *The Age of Enlightenment: Studies Presented to Theodore Besterman* (Edinburgh: Oliver and Boyd, 1967).

Barnouw, J., 'Vico and the Continuity of Science: The Relation of His Epistemology to Bacon and Hobbes', *Isis*, 71, No. 259 (1980) 609–620.

Battistini, A., *La degnità della retorica: Studi su G. B. Vico* (Pisa: Pacini, 1975).

Battistini, A. (ed.) *Nuovo Contributo alla Bibliografia Vichiana, 1971–1980* (Naples: Guida, 1983).

Battistini, A. (ed.) *Vico Oggi* (Rome: Armando Armando, 1979).

Bazzoli, B., 'Giambattista Almici e la diffusione di Pufendord nel settecento italiano', *Critica Storica*, Year 16 (1979) 3–100.

Becker, C., *The Heavenly City of the Eighteenth Century Philosophers* (New Haven: Yale University Press, 1932).

Beckett, S. *et al.*, *Our Exagmination Round his Factifaction for Incamination of Work in Progress* (Paris: Shakespeare and Co., 1929).

Bedani, G., *Vico Revisited: Orthodoxy, Naturalism and Science in the Scienza Nuova* (Oxford: Berg, 1989).

Bedani, G. L. C., 'The Poetic as an Aesthetic Category in Vico's *Scienza Nuova*', *Italian Studies*, 31 (1976) 22–36.

Bellamy, R., 'Liberalism and Historicism – History and Politics in the Thought of Benedetto Croce' (unpublished Ph.D. thesis, University of Cambridge, 1983).

Bellofiore, L., *La dottrina della provvidenza in G. B. Vico* (Padua: CEDAM, 1962).

Bellofiore, L., *Morale e storia in G. B. Vico* (Padua: CEDAM, 1972).

Benedict, R., *Patterns of Culture* (London: Routledge & Kegan Paul, 1935, 5th rpt, 1971).

Berlin, I., *Against the Current: Essays in the History of Ideas* (London: The Hogarth Press, 1979).

Berlin, I., *The Age of Enlightenment* (New York: Mentor, Houghton Mifflin, 1956, rpt. Oxford: Oxford University Press, 1979).

Berlin, I., *Vico and Herder* (London: The Hogarth Press, 1976, rpt London: Chatto & Windus, 1980).
REVIEWS:
P. Gardiner, *History and Theory*, 16, No. 1 (1977) 45–51;
A. Momigliano, 'On the Pioneer Trail', *New York Review of Books*, 23, No. 36 (Nov. 11, 1976) 33–38;
J. Morrison, 'Three Interpretations of Vico', *Journal of the History of Ideas*, 39, No. 3 (1978) 511–518.

Berlin, I., 'On the Pursuit of the Ideal', *New York Review of Books*, 35, No. 4 (March 17, 1988) 11–18.

Berry, T. M., *The Historical Theory of Giambattista Vico* (Washington, D.C.: The Catholic University of America Press, 1949).

Besaucèle, B., de, *Les Cartésians de l'Italie* (Paris: Auguste Picard, 1920).

Bianca, G., *Il concetto di poesia in Giambattista Vico* (Messina–Florence: G. D'Anna, 1967).

Blackwell, T., *An Enquiry into the Life and Writings of Homer* (London: n.p., 1735).

Blamires, C., 'Three Critiques of the French Revolution: Maistre, Bonald, and Saint-Simon' (unpublished D.Phil. thesis, University of Oxford, 1986).

Bloch, M., *Apologie pour l'histoire ou métier d'historien*, Cahiers des Annales, 3 (1949). *The Historians Craft*, P. Putnam (tr) (Manchester: Manchester University Press, 1954).

Boas, G., *Essays on Primitivism and Related Ideas in the Middle Ages* (Baltimore: Johns Hopkins University Press, 1948).

Boas, F., *General Anthropology* (Boston: Heath, 1938).

Bolletino del Centro di Studi Vichiani 1–16 (1971–1986).

Bollinger, B. L., *Aspects of Language* (New York: Harcourt, Brace & World, 1968).

Boorstin, D. J., *The Discoverers* (New York: Random House, 1983, rpt 1985).

Borsa, G., *Introduzione alla storia* (Florence: *Felice Le Monnier*, 1980).

Botton, C. A., *Church Reform in Eighteenth Century Italy* (The Hague: Martinus Nijhoff, 1969).

Bouillier, F., *Histoire de la philosophie cartésienne*, 2 vols (Paris: Durand,

1854; 3rd edn, Paris: Delagrave, 1868).

Bouwsma, W. J., 'Three Types of Historiography in Post-Renaissance Italy', *History and Theory*, 4, No. 1 (1965) 303–314.

Braudel, F., *La Méditerranée et le monde méditeranéen à l'epoque de Philippe II* (Paris: Armand Colin, 1949), *The Mediterranean and the Mediterranean World in the Age of Phillip II*, S. Reynolds (tr) 2 vols (Bungay, Suffolk: Fontana, 1972, 2nd revised tr).

Brett, G. S., *A History of Psychology*, 3 vols (London: George Allen & Unwin, 1912, 1921).

Brown, N. O., *Closing Time* (New York: Random House, 1973).

Bundy, M. W., *The Theory of Imagination in Classical and Medieval Thought*, University of Illinois Studies in Language and Literature, 12, Nos. 2–3 (1927) 1–289.

Burke, P., 'Languages and Anti-Languages in Early Modern Italy', *History Workshop Journal*, 11 (1981), 24–32.

Burke, P., *The Renaissance Sense of the Past* (London: Edward Arnold, 1969).

Burke, P., *Vico* (Oxford: Oxford University Press, 1985).

Candela, P. S., *L'unità e la religiosità del pensiero di Giambattista Vico* (Naples: Edizioni «Cenacolo Serafico», 1969).

Campbell, J., *The Mythic Image* (Princeton: Princeton University Press, 1974).

Campbell, J., *Myths, Dreams and Religion* (New York: Dutton, 1970).

Campbell, J., *Myths to Live By* (New York: Viking Press, 1972, rpt. London Souvenir Press, 1973).

Cantelli, G., *Mente Corpo Linguaggio: Saggio sull'interpretazione vichiana del mito* (Florence: Sansoni, 1986).

Caponigri, A. R., *History and Liberty: The Historical Writings of Benedetto Croce* (London: Routledge and Kegan Paul, 1955).

Caponigri, A. R., 'Philosophy and Philology: The "New Art of Criticism" of Giambattista Vico', *Modern Schoolman*, 59, No. 2 (1982) 81–116.

Caponigri, A. R., *Time and Idea: The Theory of History in Giambattista Vico* (London: Routledge and Kegan Paul, 1953).
REVIEW:
Fisch, M., *Journal of Philosophy*, 54 (1957) 648–652.

Cappello, C., *La visione della storia in G. B. Vico* (Torino: Società Editrice Internazionale, 1948).

Caraffa, N., *Gli studi giovanili e l'insegnamento accademico di G. B. Vico* (Urbino: Melchiorre Arduni, 1912).

Carpanetto, D., and G. Ricuperati, *Italy in the Age of Reason 1685–1789* (London: Longman, 1987).

Cassirer, E., *An Essay on Man: An Introduction to the Philosophy of Human Culture* (New Haven: Yale University Press, 1944).

Cassirer, E., *Language and Myth*, S. K. Langer (tr) (New York: Harper & Row, 1946).

Cassirer, E., *The Philosophy of the Enlightenment*, F. C. A. Koelln and J. P. Pettegrove (trs) (Princeton: Princeton University Press).

Cassirer, E., P. O. Kristeller, and J. H. Randall, Jr. (eds), *The Renaissance Philosophy of Man* (Chicago: University of Chicago Press, 1948).

Caverni, R., *Storia del metodo sperimentale in Italia* (Florence: G. Civelli, 1891).

Chaix-Ruy, J., *La formation de la pensée philosophique de J.-B. Vico* (Gap: Louis Jean, 1943).

Chaix-Ruy, J., *J.-B. Vico et L'Illuminisme Athée* (Paris: Mondiale, 1968).

Chaix-Ruy, J., *Vie de J.-B. Vico* (Paris: Presses Universitaires de France, 1943).

Chambliss, J. J., *Imagination and Reason in Plato, Aristotle, Vico, Rousseau, and Keats* (The Hague: Martinus Nijhoff, 1974).

Child, A., *Making and Knowing in Hobbes, Vico and Dewey, University of California Publications in Philosophy*, 16, No. 13 (1953). *Fare e conoscere in Hobbes, Vico e Dewey*, M. Donzelli (tr) (Naples: Guida, 1970).

Chorley, P., *Oil, Silk, and the Enlightenment: Economic Problems in XVIIIth Century Naples* (Naples: Istituto Italiano per gli Studi Storici, 1965).

Ciardo, M., *Le quattro epoche dello storicismo: Vico-Kant-Hegel-Croce* (Bari: Laterza, 1947).

Cipolla, C. M. (ed) *Storia dell'economia italiana* (Turin: *Boringhieri*, 1959), vol. 1, *Secoli settimo-diciassettesimo*.

Ciranna, C., *Sintesi di storia della filosofia* (Rome: Ciranna, 1986), vol. 2.

Clark, D. S. T., 'Francis Bacon: The Study of History and the Science of Man' (unpublished Ph.D. thesis, University of Cambridge, 1971).

Cochrane, E., *The Late Italian Renaissance 1525–1630* (London: Macmillan, 1970).

Cochrane, E., *Tradition and Enlightenment in the Tuscan Academies, 1690–1800* (Rome: Edizioni di storia e letteratura, 1961).

Cohen, E., 'Law, Folklore and Animal Lore', *Past and Present*, 110 (1986) 6–37.

Colie, R. L., 'Johan Huizinga and the Task of Cultural History', *American Historical Review*, 69, No. 3 (1964) 607–630.

Collingwood, R. G., *An Autobiography* (Oxford: Clarendon Press, 1939, 3rd rpt, 1982).

Collingwood, R. G., *The Historical Imagination: An Inaugural Lecture 1935* (Oxford: Oxford University Press, 1935).

* Collingwood, R. G., *The Idea of History* (Oxford: Oxford University Press, 1946, New York: Galaxy, 1956, 6th rpt).

Cortese, N., *Cultura e politica a Napoli dal Cinquecento al Settecento* (Naples: Edizioni Scientifiche Italiane, 1965).

Corsano, A. *et al.* (eds), *Omaggio a Vico* (Naples: Morano, 1968).

Corsano, A., *Umanesimo e religione in G. B. Vico* (Bari: Laterza, 1935).

Costa, G., 'An Enduring Venetian Accomplishment: The Autobiography of G. B. Vico', *Italian Quarterly*, 21, No. 79 (1980) 45–54.

Cotungo, R., *La sorte di G. B. Vico* (Bari: Laterza, 1914).

Cranston, M., *Jean-Jacques: The Early Life and Work of Jean-Jacques Rousseau, 1712–54* (London: Allen Lane, 1983, rpt. Harmondsworth: Penguin, 1987).

Crease, R. (ed.), *Vico in English* (Atlantic Highlands, New Jersey: Humanities Press, 1978).

Croce, B., *An Autobiography*, R. G. Collingwood (tr) (Oxford: Clarendon Press, 1927).

Croce, B., and F. Nicolini, *Bibliografia Vichiana* (Naples: Riccardo Ricciardi, 1947–1948).

Croce, B., 'Eternità e storicità della filosofia', *Quaderni critici*, 21 (1930).

Croce, B., *La filosofia di Giambattista Vico* (Bari: Laterza, 1911). *The Philosophy of Giambattista Vico*, R. G. Collingwood (tr) (London: Howard Latimer, 1913).

Croce, B., *Storia del Regno di Napoli* (Bari: Laterza, 1924, 1st edn, 1944, 3rd edn). *History of the Kingdom of Naples* (Chicago: University of Chicago Press, 1970).

Croce, B., *Storia della età barocca in Italia* (Bari: Laterza, 1929), vol. 23 of *Scritti di storia letteraria e politica*.

Croce, B., 'Vico', *Encyclopaedia of the Social Sciences* (New York: Macmillan, 1935) vol. 15, pp. 249–250.

Curtius, E. R., *European Literature and the Latin Middle Ages*, W. R. Trask (tr) (London: Routledge & Kegan Paul, 1953, 2nd edn, 1979).

Daedalus, 2 (1959), issue entitled 'Myth and Myth-making'.

D'Arcais, G. F., *Latinità dello storicismo vichiano* (Padua: CEDAM, 1940).

Darnton, R., *The Great Cat Massacre and Other Episodes in French Cultural History* (New York: Basic Books, 1984; 2nd edn, New York: Random House, 1985).

Davidson, D., 'On the Very Idea of a Conceptual Scheme', *Proceedings and Addresses of the American Philosophical Association*, XLVII (1973–1974) 5–20.

DeFalco, E. (ed.), *L'ideale educativo secondo le "Orationes" il "De Nostri", "l'Autobiografia" e il "Carteggio"* (Naples: De Vito, 1954).

De Maio, R., 'Muratori e il Regno di Napoli: Amicizie, Fortuna, e Polemiche', *Rivista Storica Italiana*, 85, No. 3 (1973) 756–777.

De Mas, E., *Bacone e Vico* (Torino: Edizioni di Filosofia, 1959).

De Mas, E., '*La dottrina dell'anima umana e delle sue facoltà nel sistema di Francesco Bacone*', *Filosofia*, 13 (1962) 371–408.

De Mas, E., 'On the New Method of a New Science: A Study of G. Vico', *Journal of the History of Ideas*, 32, No. 1 (1971) 85–94.

de Michelis, C., and G. Pizzamiglio (eds), *Vico e Venezia* (Florence: Leo S. Olschki, 1982).

De Rosa, G., '*Religione e società nell'Italia del settecento. I problemi della ricerca*', *Ricerche di storie sociale e religiose*, 19, No. 27 (1985) 233–246.

De Rosa, L., 'Property Rights, Institutional Change and Economic Growth in Southern Italy in the XVIIIth and XIXth Centuries', *Journal of European Economic History*, 8, No. 3 (1979) 531–552.

Descartes: Correspondance, C. Adam and G. Milhaud (eds), 8 vols (Paris: Félix Alcan, Presses Universitaires de France, 1936–1963).

Descartes, R., *The Essential Descartes*, M. D. Wilson (ed.) (New York: Mentor, 1969, rpt. Scarborough, Ontario: Meridian, 1983).

Descartes, R., *Oeuvres de Descartes*, C. Adam and P. Tannery (eds) 12 vols and supplement (Paris: Leopold Cerf, 1897–1913).

Descartes, R., *Philosophical Works of Descartes*, E. S. Haldane and G. R. T. Ross, (trs), 2 vols (Cambridge: Cambridge University Press, 1967).

Donzelli, M., *Natura e humanitas nel giovane Vico* (Naples: Istituto Italiano per gli studi storici, 1970).

de Ruggiero, G., *Myths and Ideals* (Oxford: Oxford University Press, 1946).

De Seta, C., *Storia della città di Napoli* (Rome, Bari: Laterza, 1973).

Diaz, F., 'L'Italie des Princes Eclaires', *Annales Historiques de la Révolution Française*, 51, No. 4 (1979) 581–593.

Dictionary of the History of Ideas, 5 vols (New York: Charles Scribner's Sons, 1973).

Díez-Canedo, A., *Un Estudio sobre las dos versiones de la "Ciencia Nueva" de Juan Bautista Vico* (Mexico City: Universidad Nacional Autónomade México, 1981).

Donati, B., *Nuovi studi sulla filosofia civile di G. B. Vico* (Florence: Le Monnier, 1936).

Doney, W. (ed.), *Descartes: A Collection of Critical Essays* (Garden City, New York: Doubleday [Anchor Book], 1967; rpt. London: Macmillan, 1967; rpt. Notre Dame: Notre Dame University Press, 1968).

Donoghue, D., *Imagination* (Glasgow: Glasgow University Press, 1974).

Donzelli, M. (ed.), *Contributo alla Bibliografia Vichiana (1948–1970)* (Naples: Guida, 1973).

Donzelli, M., *Natura e humanitas nel giovane Vico* (Naples: *Istituto italiano per gli studi storici*, 1970).

Dray, W., *Perspectives on History* (London: Routledge & Kegan Paul, 1980).

Dray, W. (ed.), *Philosophical Analysis and History* (New York: Harper & Row, 1966).

Dray, W., *Philosophy of History* (Englewood Cliffs, N.J.: Prentice-Hall, 1964).

Dussen, W. J. van der, *History as a Science: The Philosophy of R. G. Collingwood* (The Hague: Martinus Nijhoff, 1981).

Eagleton, Terry, *The Ideology of the Aesthetic* (Oxford: Basil Blackwell, 1990.

Eliade, M., *Myth and Reality* (London: George Allen & Unwin, 1964).

Engell, J., *The Creative Imagination: Enlightenment to Romanticism* (Cambridge, Mass.: Harvard University Press, 1981).

Fassó, G., *La lingua del Vico* (Florence: «La Nuova Italia» Editrice, 1963).

Fassó, G., *I «quattro autore» del Vico: Saggio sulla genesi della «Scienza Nuova»* (Milan: A. Giuffrè, 1949).

Fassó, G., *Vico e Grozio* (Naples: Guida, 1971).

Feldman, B. and R. D. Richardson, *The Rise of Modern Mythology, 1680–1860* (Bloomington: Indiana University Press, 1972).

Ferrari, G., *La mente di G. B. Vico* (Milan: *Società tipografica de' classici italiani*, 1837).

Ferrari, J. (G.), *Vico et L'Italie* (Paris: Eveillard, 1939).

Ferrone, V., *Scienza, natura, religione: Mondo newtoniano e cultura italiana nel primo Settecento* (Naples: Jovene, 1982).

Finetti, B., *Apologia del genere umano accusato di essere stato una volta bestia: Parte I* (Venice: Radici, 1768).

Finetti, B., *Difesa dell autoritá della scrittura contro Giambattista Vico* (Bari: Laterza, 1936, facs. of 1768).

Fisch, M., 'Vico on Roman Law', in *Essays in Political Theory Presented to G. E. Sabine*, M. R. Konvitz and A. E. Murphy (eds) (Ithaca: Cornell University Press, 1948), pp. 62–88.

Flint, R., *Vico* (Edinburgh: Blackwood, 1884).

Foreign Quarterly Review, 2 (1828), 621–661, 'Italian Literature of the Eighteenth Century'.

Foreign Review and Continental Miscellany, 5 (1830) 380–391. 'Vico – "New Science" and "Ancient Wisdom of the Italians"'.

Forum Italicum, 2, No. 4 (1968), 'Special Issue: A Homage to G. B. Vico in the Tercentenary of His Birth'.

Formigari, L., 'Language and Society in the Late Eighteenth Century', *Journal of the History of Ideas*, 35 (1974) 275-292.

France, P., *Rhetoric and Truth in France* (Oxford: Oxford University Press, 1972).

Friedrich, C. J., *The Philosophy of Law in Historical Perspective* (Chicago: University of Chicago Press, 1958).

Fubini, M. (ed.), *La cultura illuministica in Italia* (Torino: Edizione Rai Radiotelevisione Italiana, 1964).

Fubini, M., 'Letteratura scolastica', *La Cultura* (Rome: Leo S. Olschki, n.d.).

Fubini, M., *Stile e umanità di G. Vico* (Bari: Laterza, 1946).

Funkenstein, A., *Theology and the Scientific Imagination from the Middle Ages to the Seventeenth Century* (Princeton: Princeton University Press, 1986).

Gadol, E. T., 'The Idealist Foundations of Cultural Anthropology', *Journal of the History of Philosophy*, 12, No. 2 (1974) 207-225.

Gallie, W. B., *Philosophy and the Historical Understanding* (New York: Schocken, 1964).

Gardiner, P., *The Nature of Historical Explanation* (Oxford: Clarendon Press, 1952).

Gardiner, P., *The Philosophy of History* (Oxford: Oxford University Press, 1974, rpt, 1982).

Gardiner, P., *Theories of History* (New York: The Free Press, 1959).

Garin, E., 'Cartesio e l'Italia', *Giornale Critico della Filosofia Italiana*, 3rd series, Year 29, 4 (1950) 385-405.

Garin, E., *G. Pico della Mirandola* (Parma: Comitato per le celebrazioni centenarie in onore di Giovanni Pico, 1963).

Garin, E., *Giovanni Pico della Mirandola: Vita e Dottrina*, Universitá degli studi di Firenze, Facoltá di Lettere e Filosofia, 3rd series, 5.

Garin, E., *Italian Humanism: Philosophy and Civic Life in the Renaissance* (Oxford, Basil Blackwell, 1965).

Gay, P., *The Enlightenment: An Interpretation*, 2 vols (London: Weidenfeld and Nicolson, 1966).

Gentile, G., *Studi Vichiani* (Florence: Felice Le Monnier, 1927, 2nd edn).

Gentile, M. T., *Immagine e parola nella formazione dell'uomo* (Rome: Armando Armando, 1965).

Gianturco, E., 'Character, Essence, Origin and Content of the *Jus Gentium* According to Vico and Suarez', *Revue de littérature comparée*, 16 (1936) 167-172.

190   *Bibliography*

Giarrizzo, G., *Edward Gibbon e la cultura europea Settecento* (Naples: Istituto italiano per gli studi storici, 1954).

Ginzburg, C., *The Cheese and the Worms: The Cosmos of a Sixteenth-Century Miller*, John and Anne Tedeschi (trs) (Harmondsworth: Penguin, 1987, rpt.).

Giarrizzo, G., *Vico: La politica e la storia* (Naples: Guida, 1981).

Giordano, P., *Vico: Filosofo del suo tempo* (Padua, CEDAM, 1974).

Goethe, J. W., *Italian Journey*, W. H. Auden and E. Mayer (trs) (Harmondsworth: Penguin, 1982, 2nd Penguin rpt).

Golding, W., *The Inheritors* (London: Faber & Faber, 1955).

Gombrich, E. H., *Ideals and Idols: Essays on Values in History and Art* (Oxford: Phaidon Press, 1979, rpt. E. P. Dutton, 1979).

Gorman, J. L., *The Expression of Historical Knowledge* (Edinburgh: Edinburgh University Press, 1982).

Grassi, Ernesto, *Vico and Humanism: Essays on Vico, Heiddegger, and Rhetoric* (New York: Peter Lang, 1990).

Grimaldi, A. A., *The Universal Humanity of Giambattista Vico* (New York: S. F. Vannik, 1958).

Grotius, H., *De Jure Belli Ac Pacis Libri Tres*, F. W. Kelsey (tr) 2 vols (Oxford: Oxford University Press, 1925).

Habermas, J., *Theory and Practice*, J. Viertel (tr) (London: Heinemann, 1974).

Haddock, B. A., *An Introduction to Historical Thought* (London: Edward Arnold, 1980).

Haddock, B. A., 'Vico and Idealism' (unpublished D.Phil. thesis, University of Oxford, 1977).

Haddock, B. A., *Vico's Political Thought* (Swansea: Mortlake Press, 1986).

Hampshire, S., 'Vico and His "New Science"', *The Listener*, 44 (1949) 569–571.

Hampshire, S., 'Vico and Language', *New York Review of Books*, (23 February 1969) 19–22.

Hathaway, B., *The Age of Criticism: The Late Renaissance in Italy* (Ithaca: Cornell University Press, 1962).

Hazard, P., *La Crise de la conscience européenne, 1680–1715* (Paris: Boivin, 1935). *The European Mind*, J. L. May (tr) (Harmondsworth: Penguin, 1973, 3rd edn).

Hegel, G. W. F., *The Philosophy of History*, J. Sibree (tr) (New York: Dover, 1956).

Hill, C., *Intellectual Origins of the English Revolution* (Oxford: Oxford University Press, 1965, 2nd rpt, 1980).

Hughes, H. S., *Consciousness and Society* (Trowbridge and Esher: The

Harvester Press, 1979).

Humphreys, R. S., 'The Historian, His Documents, and the Elementary Modes of Historical Thought', *History and Theory*, 19, No. 4 (1980) 1–20.

Hunter, J. F. M., *Essays after Wittgenstein* (London: George Allen & Unwin, 1973).

Huppert, G., *The Idea of Perfect History* (Urbana: University of Illinois Press, 1970).

International Encyclopedia of the Social Sciences (New York: Macmillan, 1968) 'Language', vol. 9, pp. 1–22.

Jacobitti, E., *Revolutionary Humanism and Historicism in Modern Italy* (New Haven: Yale University Press, 1981).

Jacobs, M. C., *The Radical Enlightenment: Pantheists, Freemasons and Republicans* (London: George Allen & Unwin, 1981).

Jemolo, A. C., *Il giansenismo in Italia* (Bari: Laterza, 1928).

Kearns, E. J., *Ideas in Seventeenth Century France* (Manchester: Manchester University Press, 1979).

Kelley, D. R., *The Beginning of Ideology: Consciousness and Society in the French Reformation* (Cambridge: Cambridge University Press, 1981).

Kelley, D. R., *The Foundations of Modern Historical Scholarship: Language, Law and History in the French Renaissance* (New York: Columbia University Press, 1970).

Kelley, D. R., 'The Prehistory of Sociology: Montesquieu, Vico and the Legal Tradition', *Journal of the History of the Behavioral Sciences* 16, No. 2 (1980) 133–144.

Kenny, A., *Descartes: A Study of His Philosophy* (New York: Random House, 1968).

King, P. (ed.), *The History of Ideas* (Totawa, New Jersey: Barnes and Noble Books, 1983).

Koenigsberger, D., *Renaissance Man and Creative Thinking: A History of Concepts in Harmony 1400–1700* (Hassocks: The Harvester Press, 1979).

Koenigsberger, H. G., 'Decadence or Swift Changes in the Civilization of Italy and Europe in the Sixteenth and Seventeenth Centuries?' *Transactions of the Royal Historical Society*, 5th Series, 10 (1960) 1–18.

Krieger, L., *Kings and Philosophers 1689–1789* (London: Weidenfeld & Nicolson, 1970).

Kristeller, P., *Eight Philosophers of the Italian Renaissance* (Stanford: Stanford University Press, 1964).

Labanca, B., *G. Vico e i suoi critici cattolici* (Naples: Pierro, 1898).

LaCapra, D., *Rethinking Intellectual History: Texts, Contexts, Language*

(Ithaca: Cornell University Press, 1983).

Lakoff, G., and M. Johnson, *Metaphors We Live By* (Chicago: University of Chicago Press, 1980).

Land, S. K., 'The Account of Language in Vico's *Scienza Nuova*: A Critical Analysis', *Philological Quarterly*, 55 (1976) 354–373.

Lanza, F., *Saggi di poetica vichiana* (Varese: Magenta, 1961).

Langer, S. K., *Philosophy in a New Key* (Cambridge, Mass.: Harvard University Press, 1942, 2nd edn, 15th rpt. New York: Mentor, 1951).

Lee, D. E., and R. N. Beck, 'The Meaning of "Historicism"', *American Historical Review*, 59, No. 3 (1959) 568–577.

Lefèvre, R., *L'Humanisme de Descartes* (Paris: Presses Universitaires de France, 1957).

Lepschy, G., '*Fantasia e immaginazione*', *Lettere Italiane*, 39, No. 1 (1987) 20–34.

Lévi-Strauss, C., *Myth and Meaning* (London: Routledge & Kegan Paul, 1978).

Lévi-Strauss, C., *La pensée sauvage* (Paris: Plon, 1962).

Lévi-Strauss, C., *The Scope of Anthropology* (London: Jonathan Cape, 1967).

Lévi-Strauss, C., *Tristes tropiques* (Paris: Plon, 1955).

Lilla, Mark, 'A Preface to Vico: Skepticism, Politics, Theodicy' (unpublished Ph.D. thesis, Harvard University, 1990).

Lion, A., *The Idealist Conception of Religion: Vico, Hegel, Gentile* (Oxford, Clarendon Press, 1932).

Longo, M., *Giambattista Vico* (Turin: Bocca, 1921).

Löwith, K., *Meaning in History: The Theological Implications of the Philosophy of History* (Chicago: University of Chicago Press, 1949).

Löwith, K., *Vicos Grundsatz: verum et factum convertuntur* (Heidelberg: Carl Winter Universitätsverlag, 1968).

Malinowski, B., *Magic, Science, and Religion and Other Essays* (Garden City, N.Y.: Doubleday, 1948).

Manno, A. G., *Lo storicismo di G. B. Vico* (Naples: Istituto Editoriale del Mezzogiorno, 1965).

Manson, R., *The Theory of Knowledge of Giambattista Vico* (Hamden, Connecticut: Archon Book, 1969).

Manuel, F., *The Eighteenth Century Confronts the Gods* (Cambridge, Mass.: Harvard University Press, 1959).

Marini, L., *Il Mezzogiorno d'Italia di fronte a Vienna e a Roma* (Bologna: Pátron, 1970).

Marini, L., *Pietro Giannone e il Giannonismo a Napoli nel Settecento* (Bari: Laterza, 1950).

Martin-Trigona, H. V., *Logical Proof and Imaginative Reason in Selected Speeches of Francis Bacon* (unpublished Ph.D. dissertation, University of Illinois, 1967).

Marx, K., *The Eighteenth Brumaire of Louis Bonaparte* (New York: International Publishing Company, 1898). *The Eighteenth Brumaire of Louis Bonaparte* (tr n.g.) (Peking: Foreign Languages Press, 1978). *Der achtzehnte Brumaire des Louis Bonaparte* (Hamburg: 1885).

Mastellone, S., *Pensiero politico e vita culturale a Napoli nella seconda metà del seicento* (Messina-Florence: G. D'Ana, 1965).

Maugain, G., *Étude sur l'évolution intellectuelle de l'Italie de 1657 à 1750 environ* (Paris: Librarie Hachette, 1909).

Mazlish, B., *The Riddle of History: The Great Speculators from Vico to Freud* (New York: Harper and Row, 1966).

Mazzeo, J., *Renaissance and Seventeenth Century Studies* (New York: Columbia University Press, 1964).

Megill, A., 'Aesthetic Theory and Historical Consciousness in the Eighteenth Century', *History and Theory*, 17, No. 1 (1978) 29–62.

Meek, R. L., *Social Science and the Ignoble Savage* (Cambridge: Cambridge University Press, 1976).

Merleau-Ponty, M., *La Prose du monde* (Mayenne: Gallimard, 1969). *The Prose of the World*, J. O'Neill (tr) (Evanston, Illinois: Northwestern University Press, 1973).

Migliorini, B., *La lingua italiana* (Florence: Sansoni, 1960). Revised by T. G. Griffith as *The Italian Language* (London: Faber, 1984, 2nd English edn).

Mill, J. S., *Dissertations and Discussions*, 2 vols in 4 (London: John W. Parker and Son, 1859–1875).

Mills, W. J., 'Positivism Reversed: The Relevance of Giambattista Vico', *Transactions of the Institute of British Geographers*, N.S. 7, Nos. 1–14 (1982) 1–14.

RESPONSE:

D. Kunze, 'Giambattista Vico: As a Philosopher of Place, Comments on the Recent Article by Mills', *Transactions of the Institute of British Georgraphers*, N.S. 8 (1983) 237–248.

Mink, L. O., *Mind, fhistory and Dialectic: The Philosophy of R. G. Collingwood* (Bloomington: Indiana University Press, 1969).

Modica, G., *La filosofia del «senso comune» in G. Vico* (Caltanissetta-Rome: Sciascia, 1983).

Modica, G., 'Sul ruolo del «senso comune» nel giovane Vico', *Rivista Filosofia Neo-Scolastica*, 75 (1983), 243–262.

Momigliano, A., 'Gibbon from an Italian Point of View', *Daedalus*, 105, No. 3 (1976) 125–135.

Momigliano, A., 'The One True History', *Times Literary Supplement*, (5 September 1975) 982–983.

Momigliano, A., *Studies in Historiography* (New York: Harper and Row, 1966).

Momigliano, A., 'Vico's *Scienza Nuova*: Roman "*Bestioni*" and Roman "*Eroi*", *History and Theory*, 5, No. 1 (1966) 3–23.

Moncallero, G. L., *Teorica d'Arcadia* (Florence: Olschki, 1953), vol. 1 of *L'Arcadia*.

Mondolfo, R., *Il 'verum-factum' prima di Vico* (Naples: Guida, 1969).

Montano, R., 'Vico's Opposition to the Enlightenment', *Italian Quarterly*, 17, No. 68 (1974) 3–34.

Monti, S., *Sulla tradizione e sul testo delle orazioni inaugurali di Vico* (Naples: Guida, 1977).

Mooney, M., *Vico in the Tradition of Rhetoric* (Princeton: Princeton University Press, 1985).

Moravia, S., *La scienza dell'uomo nel Settecento* (Bari: Laterza, 1970).

Morrison, J., 'Vico and Spinoza', *Journal of the History of Ideas* 41, No. 1 (1980) 49–68.

Morrison, J., 'Vico's Doctrine of the Natural Law of the Gentes', *Journal of the History of Philosophy*, 16 (1978) 47–60.

Morrison, J. 'Vico's Principle of *Verum* is *Factum* and and the Problem of Historicism', *Journal of the History of Ideas*, 39, No. 4 (1978) 579–595.

New Vico Studies, 1–7 (1983–1989).

Nicolini, F., *Commento storico alla seconda Scienza Nuova*, 2 vols (Rome: Edizioni di «Storia Letteratura», 1949).

Nicolini, F., *La giovinezza di Giambattista Vico (1668–1700)* (Bari: Laterza, 1932).

Nicolini, F., 'Jean-Baptiste Vico dans l'histoire de la pensèe', *Cahiers d'histoire mondiale* 7, No. 2 (1963) 299–319.

Nicolini, F., *Giambattista Vico, Opere* (Milan, Naples: Riccardo Ricciardi, 1953).

Nicolini, F., *La religiosità di Giambattista Vico* (Bari: Laterza, 1949).

Nicolini, F., *Sugli studi omerici di Giambattista Vico* (Rome: Accademia Nazionale dei Lincei, 1954).

Nicolini, F., *Vico Storico* (Naples: Morano, 1967).

Noether, E. P., *Seeds of Italian Nationalism, 1700–1815* (New York: Columbia University Press, 1951).

Nuzzo, E., *Vico* (Florence: Vallecchi, 1974).

O'Brien, G. D., *Hegel on Reason and History* (Chicago: University of Chicago Press, 1975).

Opinioni e Giudizi di Alcuni Illustri Italiani e Stranieri sulle Opere di Giambattista Vico (Naples: Iovene, 1863) [Library of Congress, Law Library – LAW, Italy, 7, OPIN]

Ottavanio, C., *Metafisica dell'essere parziale* (Naples: Rondinella, 1954).

Otto, S., *Materialien zur Theorie der Geistesgeschichte* (Munich: Wilhelm Fink Verlag, 1979).

Paci, E., *Ingens silva* (Milan: Mondadori, 1949).

Pandolfi, C., 'Appunti di filologia vichiana', *Giornale Italiano di Filologia*, N.S. 8 (29), No. 2 (1977) 181–194.

Papini, M., *Arbor humanae linguae* (Bologna: Capelli, 1984).

Parekh, B. (ed.), *Jeremy Bentham: Ten Critical Essays* (London: Frank Cass, 1974).

Pérez-Ramos, Antonio, *Francis Bacon's Idea of Science and the Maker's Knowledge Tradition* (Oxford: Clarendon Press, 1988).

Perrotta, P. C., 'Giambattista Vico, Philosopher-Historian', *Catholic Historical Review*, 20 (1934–5) 384–410.

Pico della Mirandola, G. F., *On the Imagination*, H. Caplan (tr) (New Haven: Yale University Press, 1930).

Piovani, P., 'Philosophie et histoire des idées', *Revue européenne des sciences sociales*, 10, No. 28 (1972) 5–245.

Plato: The Collected Dialogues, E. Hamilton and H. Cairns (eds) (Princeton: Princeton University Press, 1982, 11th pr.).

Pomeau, R., *L'Europe des lumières* (Pomeau: Stock, 1966).

Pompa, L., *A Study of the 'New Science'* (Cambridge: Cambridge University Press, 1975, 2nd edn, 1990).

Pompa, L., and W. Dray (eds), *Substance and Form in History* (Edinburgh: University of Edinburgh Press, 1981).

Pompa, L., 'Vico's Science', *History and Theory*, 10 (1971) 49–83.

Pompa, L., 'Vico's Theory of the Causes of Historical Change' (Tunbridge Wells, Kent: *Institute for Cultural Research*, 1971, rpt. 1979).

Pons, A., 'Nature et histoire chez G. B. Vico', *Les études philosophiques*, No. 216 (1961) 39–53.

Popper, K. R., *The Poverty of Historicism* (London: Routledge & Kegan Paul, 1957).

Porter, R., and M. Teich, *The Enlightenment in National Context* (Cambridge: Cambridge University Press, 1981).

Prezzolini, G., *The Legacy of Italy* (New York: Vanni, 1948).

Procacci, G., *History of the Italian People* (London: Weidenfeld and Nicolson, 1968).

Quigley, H., *Italy and the Rise of a New School of Criticism in the Eighteenth Century* (Perth: Munro & Scott, 1921).

Quint, David, *Origin and Originality in Renaissance Literature: Versions of the Source* (New Haven: Yale University Press, 1983).

Rak, M., *Letture vichiane* (Naples: Liguori, 1971).

Renaldo, J. J., 'Bacon's Empiricism, Boyle's Science, and the Jesuit Response in Italy', *Journal of the History of Ideas*, 37, No. 4 (1976) 689–695.

Ricoeur, P., *Hermeneutics and the Human Sciences: Essays on Language, Action, and Interpretation* (Cambridge: Cambridge University Press, 1981).

Ricoeur, P., *Interpretation Theory* (Fort Worth: Texas Christian University Press, 1976).

Ricoeur, P., *The Reality of the Historical Past* (Milwaukee: Marquette University Press, 1984).

Rieff, P., *The Mind of the Moralist* (London: Victor Gollancz, 1960).

Rigault, M. H., *Histoire de la querelle des anciens et des moderns* (Paris: Hachette, 1856).

Robertson, J. C., *Studies in the Genesis of the Romantic Theory in the Eighteenth Century* (Cambridge: Cambridge University Press, 1923).

Rorty, R., J. B. Schneewind and Q. Skinner (eds), *Philosophy in History* (Cambridge: Cambridge University Press, 1985, 2nd rpt).

Rosa, M., *Cattolicesismo e lumi nel settecento italiano* (Rome: Herder, 1981).

Rosa, M., *Dispotismo e libertà nel Settecento: interpretazioni «repubblicane» di Machiavelli* (Bari: Dedalo Litostampa, 1964).

Rossi, P. *Francesco Bacone: Dalla magica all scienza* (Bari: Laterza, 1957).

Rossi, P., *'Gli studi vichiani' Immagini del Settecento in Italia*, (Bari: Laterza, 1980) 98–107.

Rossi, P., *I segni del tempo: Storia della terra e storia delle nazioni da Hooke a Vico* (Milan: Feltrinelli, 1979).

Rossi, P., *Le sterminate antichità: Studi vichiani* (Pisa: Nistri-Lischi, 1969).

Rossi, P., 'Lineamenti di storia della critica vichiana', in W. Binni (ed.), *I classici italiani nelle storia della critica* (Florence: La nuova Italia, 1961) vol. II.

Roy, J. H., *L'imagination selon Descartes* (Paris: Gallimard, 1944).

Runes, D. D. (ed.), *The Dictionary of Philosophy* (London: George Routledge & Sons, 1944).

Rudnick, S. D., 'From Created to Creator: Conceptions of Nature and Authority in Sixteenth Century England' (unpublished Ph.D. dissertation, Brandeis University, 1963).

Rudnick Luft, S. D., 'A Genetic Interpretation of Divine Providence in Vico's *New Science*', *Journal of the History of Philosophy*, 20 (1982), 151–169.

Russell, B., *The Analysis of Mind* (London: George Allen & Unwin, 1921).

Ryder, A., *The Kingdom of Naples under Alfonso the Magnanimous: The Making of a Modern State* (Oxford: Clarendon Press, 1976).

Ryle, G., *The Concept of Mind* (London: Hutchinson University Library, Senior Series, 1949).

Ryle, G., *Philosophical Arguments: An Inaugural Lecture 1945* (Oxford: Clarendon Press, 1945).

Saccenti, M., *Lucrezio in Toscana* (Florence: Olschki, 1966).

Saggi e Ricerche sul Settecento (Naples: Istituto Italiano per gli Studi Storici, 1968).

Said, E., *Beginnings: Intention and Method* (New York: Basic Books, 1975).

Said, E., *The World, The Text and the Critic* (Cambridge, Massachusetts: Harvard University Press, 1983).

Sanna, M., *Catalogo Vichiano Napoletano* (Naples: Bibliopolis, 1986).

Santinello, G., *Storia delle storie generali della filosofia* (Brescia: La Scuola, 1979), vol. II, *Dell'éta Cartesiana a Brucker*.

Sapir, E., *Culture, Language and Personality* (Berkeley: University of California Press, 1956).

Sartre, J.-P. *L'imaginaire* (Paris: Gallimard, 1940). *Imagination: A Psychological Critique* (Ann Arbor: University of Michigan Press, 1972).

Schaeffer, J. D., 'Vico's Rhetorical Model of the Mind: *Sensus Communis* in the *De nostri temporis studiorum ratione*', *Philosophy and Rhetoric*, 14 (1981) 152–167.

Schenk, H. G., *The Mind of the European Romantics: An Essay in Cultural History* (London: Constable, 1966; Oxford: Oxford University Press, 1979, 2nd edn).

Schmitt, C. and Skinner, Q. (eds), *The Cambridge History of Renaissance Philosophy* (Cambridge: Cambridge University Press, 1988).

Shorter, J. M., 'Imagination', *Mind*, N.S. 61 (1952) 528–542.

Siedentop, L., 'The Limits of Enlightenment: A Study of Conservative Social and Political Thought in Early Nineteenth France, with reference to Maine de Biran and Joseph de Maistre' (unpublished D. Phil. thesis, Oxford, 1966).

Simioni, A., *Le Origini del Risorgimento Politico dell'Italia Meridionale*, 2 vols (Messina-Rome: Giuseppee Principato, 1925).

Simonsuuri, K., *Homer's Original Genius: Eighteenth-Century Notions of the Early Greek Epic (1688–1798)* (Cambridge: Cambridge University Press, 1979).

Skinner, Q., *The Foundations of Modern Political Thought*, 2 vols (Cambridge: Cambridge University Press, 1978, 2nd rpt. 1980).

Smalley, W. A., *Readings in Missionary Anthropology II* (South Pasadena, California: William Carey Library, 1978).

Sorrentino, A., *La retorica e la poetica di G. B. Vico* (Turin: Bocca, 1927).

Spaventa, B., *La filosofia italiana* (Bari: Laterza, 1908).

Sprigge, C., *Benedetto Croce: Man and Thinker* (Cambridge: Bowes and Bowes), 1952.

Steiner, G., *After Babel* (Oxford: Oxford University Press, 1976).

Steinberg, M., 'The Twelve Tables and their Origin: An Eighteenth Century Debate', *Journal of the History of Ideas*, 43, No. 3 (1982) 379–396.

Stern, F., *The Varieties of History from Voltaire to the Present* (London: Meridian Books, Thames and Hudson, 1957).

Stewart, K., 'Ancient Poetry as History in the Eighteenth Century', *Journal of the History of Ideas*, 3 (1958) 335–347.

Stoye, J. W., *English Travellers Abroad, 1604–1667* (London: Jonathan Cape, 1952).

Stromberg, R. N., 'History in the Eighteenth Century', *Journal of the History of Ideas*, 12, No. 3 (1951) 295–304.

Tagliacozzo, G., D. P. Verene, V. Rumble (eds), *A Bibliography of Vico in English 1884–1984* (Bowling Green, Ohio: Philosophy Documentation Center, 1986).

Tagliacozzo, G. and H. White (eds), *Giambattista Vico: An International Symposium*, (Baltimore: The Johns Hopkins Press, 1969).

Tagliacozzo, G., and D. P. Verene (eds), *Giambattista Vico's Science of Humanity* (Baltimore: Johns Hopkins University Press, 1976).

REVIEW:

J. Barnouw, *Eighteenth Century Studies*, 10 (1977) 384–388.

Tagliacozzo, G., M. Mooney and D. P. Verene (eds), *Vico: Past and Present* (Atlantic Highlands, New Jersey: Humanities Press, 1981).

Tagliacozzo, G., M. Mooney, and D. P. Verene (eds), *Vico and Contemporary Thought* (Atlantic Highlands, New Jersey: Humanities Press, 1979, rpt London: Macmillan, 1980), rpt. from *Social Research*, 43, Nos. 3–4 (1976).

Tagliacozzo, G. (ed.), *Vico and Marx: Affinities and Contrasts* (Atlantic Highlands, New Jersey: Humanities Press, 1983; London: Macmillan Press, 1983).

Tessitore, F. (ed.), *Giambattista Vico nel terzo centenario della nascita* (Salerno: *Istituto Universitario di Salerno*, 1968).

Titone, V., *La storiografia dell'illuminismo in Italia* (Milan: Mursia, 1969, first edition 1952).

Todd, J. and J. Cono, 'Vico and Collingwood on "The Conceit of the Scholars"', *History of European Ideas*, 6, No. 1 (1983) 59–69.

Torrini, M., 'L'Accademia degli Investiganti, Napoli, 1663–1670', *Quaderni Storici*, 48 (1981) 845–883.

Torrini, M., 'Il problema del rapporto scienza-filosofia nel pensiero del primo Vico', *Physis*, 20, Nos. 1–4 (1978) 103–121.

Ursini-Scuderi, S., G. B. *Vico come fondatore della sociologia moderna*, (Palermo: Giuseppe Pedone Lauriel, 1888).

Vasoli, C., *L'enciclopedismo del Seicento* (Naples: Bibliopolis, 1978).

Vaughan, C. E., *Giambattista Vico: An Eighteenth Century Pioneer, The Bulletin of the John Rylands Library*, 6, No. 3 (July 1921). Reprinted for private circulation: Aberdeen: Aberdeen University Press, 1921).

Vaughan, C. E., *Studies in the History of Political Thought Before and and After Rousseau*, 2 vols (Manchester: Manchester University Press, 1925).

Vaughan, F., *The Political Philosophy of Giambattista Vico* (The Hague: Martinus Nijhoff, 1972).

Venturi, F., 'Alle origini dell'Illuminismo Napoletano: Dal carteggio di Bartolomeo Intieri', *Rivista Storica Italiana*, 71 (1959) 416–456.

Venturi, F., *Italy and the Enlightenment*, S. Corsi (tr) (London: Longman, 1972).

Verene, D. P., *Vico's Science of Imagination* (Ithaca: Cornell University Press, 1981).

REVIEWS:

Alberti, A., *The Journal of Modern History*, 55 (1983) 151–152.

Armour, L., *Library Journal* 106 (1981) 887.

Bevilacqua, V. M., *Quarterly Journal of Speech*, 69 (1983) 444–447.

Bibliographical Bulletin of Philosophy, 29 (1982) 112.

Blasi, A., *Journal of the Behavioral Sciences*, 19 (1983) 265–266.

Cain, S., *Religious Studies Review*, 8 (1982) 162.

Caponigri, A. R., *The Modern Schoolman*, 60 (1983) 221–224.

Choice, 19 (1981) 226.

Dupree, D., *The Review of Metaphysics*, 35 (1982) 916–917.

Evangeliou, C., *Philosophia*, 12 (1982) 445–447.

Haddock, B. A., *Religious Studies*, 19 (1983) 549–552.

Jung, H. Y., 'Vico's Rhetoric: A Note on Verene's *Vico's Science of Imagination*', *Philosophy and Rhetoric*, 15, No. 3 (1982) 187–202.

Lovekin, D., *Philosophy and Rhetoric*, 16 (1983) 55–60.

Milbank, J., *History of European Ideas*, 4 (1983) 337–342.

Munk, A., *Journal of Philosophy and Social Science*, (1984) 356–357.

200 Bibliography

Pompa, L., *International Studies in Philosophy*, 17, No. 1 (1985) 101–103.

Psychological Medicine, 12 (1982).

Steven, R. S., *Ethics*, 92 (1982) 792.

Strong, E. F., *Journal of the History of Philosophy*, 21 (1983) 273–275.

Times Literary Supplement, (Nov. 6, 1981), 1309.

Walsh, W. H., *British Journal of Aesthetics*, 22 (1982) 378–380.

Verri, A., G. B. *Vico nella cultura contemporanea* (Lece: Milella, 1979).

Villari, R., 'La Spagna, L'Italia, e L'Assolutismo', *Studi Storici*, 18, No. 4 (1977) 5–22.

Villa, G., *La filosofia del mito* (Milan: Bocca, 1949).

Vitale, M., 'Norma linguisitica cruscante in scrittori atomisti e libertini del secondo seicento napoletano', *Società nazionale di scienze, lettere e arti in Napoli*, n.d.

Vittorini, D., 'Giambattista Vico and Reality: an Evaluation of the "De nostri temporis studiorum ratione" (1708)', *Modern Language Quarterly*, 13 (1952) 90–98.

von Otto, S., '*Imagination und Geometrie: Die Idee kreativer Synthesis*', *Archiv für geschichte der philosphie*, 63, No. 3 (1981).

Wainwright, E. H., 'The Historical Thought of Giambattista Vico', *Kleio*, 9, Nos. 1–2 (1977) 1–21.

Wallace, K. R., *Francis Bacon on the Nature of Man* (Urbana: University of Illinois Press, 1967).

Walsh, W. H., *An Introduction to Philosophy of History* (London: Hutchinson University Library, 1951, 3rd edn, 1967).

Warnock, M., *Imagination* (London: Faber and Faber, 1976).

Wasserman, E. R. (ed.), *Aspects of the Eighteenth Century* (Baltimore: Johns Hopkins University Press, 1965).

Weinberg, B., *A History of Literary Criticism in the Italian Renaissance*, 2 vols (Chicago: University of Chicago Press, 1961).

Weinryb, E., 'Re-enactment in Retrospect', *The Monist*, 72, No. 4 (1989) 568–580.

Whitman, C. H., *Homer and the Heroic Tradition* (Cambridge, Mass.: Harvard University Press, 1965).

Whittaker, T., 'Vico's New Science of Humanity', *Mind*, 35 (1926) 59–71, 204–221, 319–336, rpt. in *Reason: A Philosophical Essay, with Historical Illustrations – Comte, Mill, Schopenhauer, Vico, Spinoza* (Cambridge: Cambridge University Press, 1934).

Wohlfart, G., *Denken der Sprache: Sprache und Kunst bei Vico, Hamann, Humboldt und Hegel* (Freiburg/Munich: Verlag Karl Alber, 1984).

Wolf, E. R., *Europe and the People Without History* (Berkeley: University of California Press, 1982).
bibliography

Woolf, S., *A History of Italy* (London: Methuen, 1979, rpt. 1986).

Yates, F., *The Art of Memory* (London: Routledge & Kegan Paul, 1966, rpt 1984).

Zagorin, P., 'Vico's Theory of Knowledge', *The Philosophical Quarterly*, 34, No. 134 (1984) 15–30.

RESPONSE:

Berlin, I., 'Discussions on Vico', *Philosophical Quarterly* 35, No. 140 (1985) 281–290.

Zagorin, P., 'Berlin on Vico', *Philosophical Quarterly*, 35, No. 140 (1985) 290–296.

Zolla, E., *The Uses of Imagination and the Decline of the West* (Ipswich: Golgonooza Press, 1978; rpt. from *Sophia Perennis*, 1, No. 1).

Index

CPSIA information can be obtained
at www.ICGtesting.com
Printed in the USA
LVOW07*0304211117
557037LV00008B/128/P